Laboratory Exercises in Biological Diversity
Protists, Plants, Fungi and Animals

By Alan R. Holyoak, PhD

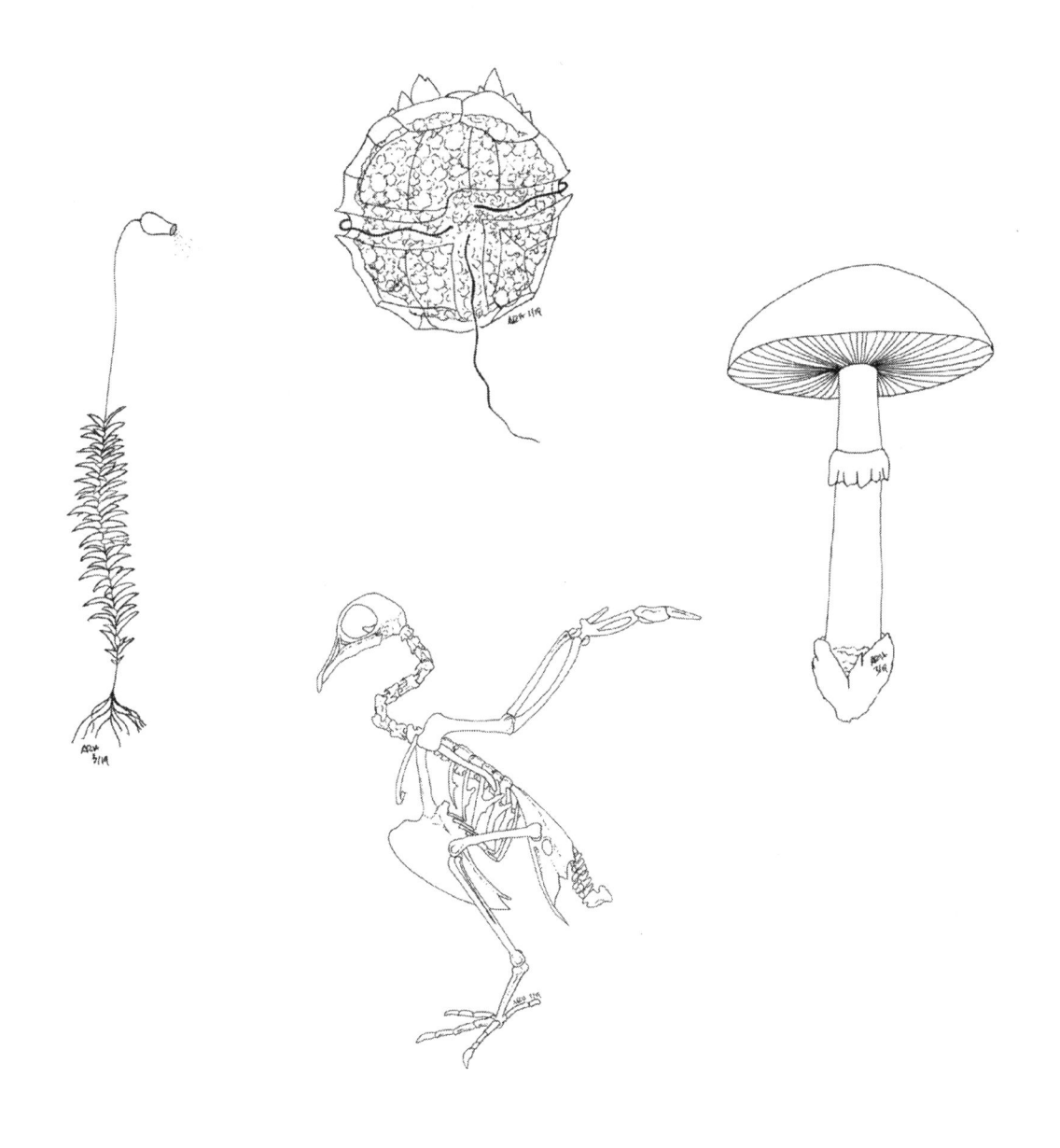

LABORATORY EXERCISES IN BIOLOGICAL DIVERSITY: THE EUKARYOTES PROTISTS, PLANTS, FUNGI AND ANIMALS, FIRST EDITION

Amazon Kindle Direct Publishing. Copyright © 2023 by Alan R. Holyoak.

All rights reserved, including text and illustrations, except as otherwise indicated.

Identifier: **ISBN: 9798866902934**

Imprint: Independently Published

No part of this publication may be reproduced or distributed in any form or by any means.

The author retains the right to make this work available for internal distribution at Brigham Young University – Idaho

Acknowledgement: The chapters of this laboratory manual that focus on plants and fungi are based on materials provided by Dr. Gary Baird, Department of Biology, BYU-Idaho. I thank him for his expert feedback and support.

DEDICATION

In memory of Kat, my best supporter, partner and friend.

Table of Contents

- Introduction 4
- Chapter 1: Phylogenetic Analysis and Introduction to Laboratory Work 5
- Chapter 2: Introduction to Microscopy and Plant and Animal Cells 15
- Chapter 3: Protists 29
- Chapter 4: Plantae - Nonvascular Plants 40
- Chapter 5: Plantae - Non-flowering Vascular Plants 50
- Chapter 6: Plantae - Flowering Vascular Plants 67
- Chapter 7: Fungi 81
- Chapter 8: Animalia - Basal Animals 93
- Chapter 9: Animalia - Lophotrochozoa 104
- Chapter 10: Animalia - Ecdysozoa 118
- Chapter 11: Animalia – Deuterostomes 141
- Chapter 12: Animalia – Vertebrates 149
- Index 173

Introduction

For hundreds of years, scientists have worked to describe and classify the diversity of life. As a result of these efforts, nearly two million species of living things have been described so far. And current estimates suggest that there may be more than 30 million species on the planet!

The focus of this lab manual is the to provide you with hands-on experiences that will help you become familiar with this diversity. These exercises center on the body plans of major groups of eukaryotic life: protists, plants, fungi and animals.

This lab manual is can fill a variety of roles. It includes enough to depth to support a laboratory experience for a one-semester course in biological diversity for biology majors. It can also be used to support a general education course for non-majors, or a course for education majors who need this information to prepare for the PRAXIS exam.

Lastly, throughout my career, I watched as the price of commercially available laboratory manuals skyrocketed, now often costing students more than $100.00/copy. So, in an effort to help ease the financial burden of lab work on my students, I started producing my own lab materials. This lab manual grew of these materials. And it is now my pleasure to offer this set of laboratory exercises at a fraction of the cost of the lab manuals currently available from textbook publishing companies.

Welcome to the world of biological diversity!

Chapter 1: Phylogenetic Analysis and Introduction to Laboratory Work

Scientists use **phylogenetic analysis**, also known as **cladistics**, to test hypotheses about taxonomic groupings of organisms. Historically, classification was carried out largely by analysis of shared anatomical or developmental traits, regardless of whether these traits were ancestral or derived. The more traits two taxa shared in common, the more closely related they were believed to be. This approach worked fairly well overall, but with the development of molecular methods we can now use molecular traits such as DNA and protein amino acid sequencing along with traditional anatomical and developmental traits in our analyses. This improved approach revealed a need to reconsider some of our long-accepted groupings of taxa.

Historical methods of classification used both ancestral and derived traits, but modern phylogenetics uses only derived traits; traits that a taxon has that its ancestral taxon did not have. When a group of taxa share the same derived traits compared to a common ancestor, we consider them to be related to each other. This updated approach to classification has led to some surprising changes in classification. For example, in the past, we concluded that members of phylum Nematoda, the roundworms, had anatomically simple bodies, so we placed them near the base of the tree of animal life. Recent analysis of molecular, biochemical and anatomical characteristics of roundworms, however, resulted in the unexpected conclusion that they are not a basal taxon at all, but are surprisingly more closely related to arthropods than they are to other basal taxa.

The goal of this exercise is to introduce you to the basics of cladistics, and to give you a chance to work through an exercise where you work as part of a team to produce your own phylogenetic tree. But before you do this, you need to become familiar with the vocabulary of cladistics.

Task #1: Learn the Following Terms of Cladistics

> **Taxon** – Any scientifically named group of organisms; a taxon can be as small as a species or as large as a domain.
>
> **Ancestral taxon** – A taxon that gives rise to at least one descendant taxon by speciation.
>
> **Daughter taxon** – A taxon that descended from an ancestral taxon by speciation.
>
> **Monophyletic group or Clade** – A group of taxa that includes an ancestral taxon and all of its descendant taxa. The identification or creation of monophyletic taxa is the goal of cladistics.
>
> **Paraphyletic taxon** – A violation of monophyletic groupings because at least one daughter taxon of an ancestral taxon is not included in the grouping. One goal of cladistics is to identify and replace paraphyletic taxa with monophyletic groupings.
>
> **Polyphyletic taxon** – A violation of monophyletic groupings because more than one ancestral taxon is required to describe the origins of the taxa within the group. These taxa, when discovered are rejected and replaced with monophyletic groupings.
>
> **Ingroup** – The group of taxa you are investigating.

Sister taxon – The taxon that is most closely related to the ingroup, but it is not part of the ingroup.

Outgroup – A taxon, ideally the sister taxon, that is included in phylogenetic analysis for comparative purposes. The outgroup will have only ancestral characteristics and will always be the first taxon branching off of the cladogram being created.

Character/Trait – An anatomical, developmental, molecular or biochemical trait of a taxon. A character can be gained and then lost during evolution, but the probability of that exact same character being lost and then regained is so low that we do not consider it to be a possibility during analysis.

Symplesiomorphy – A shared ancestral trait, i.e., a trait a taxon has because it inherited it from its ancestor. Symplesiomorphies are not included or used in cladistics.

Synapomorphy – A shared derived trait, i.e., a trait a taxon has that its ancestor did not have. Only synapomorphies are used in cladistics.

Character polarity – The relationship between different versions of a particular character. The outgroup helps us determine which version of a character state is ancestral and which is derived.

Cladogram or Phylogenetic tree – A diagram that represents the evolutionary history of a group of taxa as well as relatedness between taxa on the tree by its branching pattern and the locations of taxa at the tips of branches of the tree.

Task #2: Produce a cladogram

The goal of cladistics is to describe the evolutionary relatedness between taxa and to resolve problems of paraphyletic and polyphyletic taxa so that all taxa are monophyletic. Cladistics includes the practice of producing cladograms, also known as phylogenetic trees based on parsimony. Though not definitively correct, the most parsimonious (i.e., most likely) tree is the one that has the fewest number of character changes in it.

Professional taxonomists do not do this work by hand, instead they use computer programs to generate cladograms that they then analyze. In this exercise however, you will work as a member of a team to produce a cladogram by hand. This will give you insights into how phylogenetic analysis is done and how cladograms are produced.

1) Become familiar with procedures used to produce cladograms
    a. Identify the ingroup, outgroup and characters to use in the analysis.
    b. Produce a matrix of characters and taxa
        i. Write the names of the outgroup taxon and ingroup taxa along the top of the matrix.
        ii. Write the names of characters down the side of the matrix.
    b. Find out which characters each taxon has and enter an appropriate notation in each cell for each taxon and character combination.

i. This is relatively easy as long as a trait has only two options: present or absent. If a character is present in a taxon, write "1" in the cell. If it is absent write "0" in the cell.
ii. Things get a bit trickier when there are multiple states for a character. For example, there are multiple options for body symmetry. These include asymmetrical, radial, biradial and bilateral. We assign these the numbers 1, 2, 3 and 4, respectively, for the character. These numbers represent the character polarity from ancestral to derived character states.
iii. The order in which you complete the analysis varies between workers, but the matrix should be completed before the tree is generated.
c. A cladogram is complete when all taxa and character combinations in the matrix are accounted for on the cladogram.
d. The cladogram can be drawn as a long straight line called the backbone with side branches that come off the backbone. Characters are indicated either on the backbone or on the side branches. Taxa are found only at tips of the branches. The easiest way to explain how to do cladistic analysis is through an example.

2) Work through this example of how to generate a cladogram

Taxa and characters:
   Outgroup: Prokaryotes (All prokaryotes are lumped together for convenience.)

   Ingroup: Eukarya – Animalia, Fungi, Plantae and Protista (Protista is no longer a viable taxon but it is used here for convenience.)

   Characters:
   - DNA
   - Nucleus
   - Chloroplasts (present in at least some members of the taxon)
   - *Hox* genes
   - Cell walls (not peptidoglycan)
   - Strict multicellularity (all species in a taxon must be multicellular during at least some stage of the life cycle).

   List of taxa and their characters from the list above:
   - Prokaryotes: DNA
   - Animalia: DNA, Nucleus, *Hox* genes, Strict multicellularity
   - Fungi: DNA, Nucleus, Cell Wall, Strict multicellularity
   - Plantae: DNA, Nucleus, Chloroplasts, Cell Walls, Strict multicellularity
   - Protista: DNA, Nucleus, Chloroplasts, Cell Wall

   a. Draw a matrix, i.e., a table with a column for each taxon and a row for each character. Write a "1" in each cell of the table where a taxon has a particular trait and write a "0" in the cell where it does not. Once your matrix of taxa and characters is complete you are ready to generate a cladogram.

b. Draw a long diagonal line. This is the backbone of the cladogram.
    i. Any character that appears on the backbone is found in all taxa farther up the tree.
    ii. Any character that appears on a side branch applies only to the taxon/taxa on that side branch.
b. Start by drawing a short hash mark across the backbone near its base and write "DNA" next to it. All taxa have this trait – it is actually a symplesiomorphy but we use it as a characteristic of the outgroup. This mark at the base of the tree indicates that all taxa on the tree have this trait – also as indicated in your matrix by the complete row of ones next to the character DNA.
c. By definition the outgroup branches off first. If you look at your completed matrix you will see that the only character Prokaryotes have is DNA. Draw a line extending perpendicularly to the backbone but above the hash mark for DNA and write "Prokaryotes" at its tip. The taxon Prokaryotes and all of its characters we included in the matrix are now accounted for.
d. Decide which taxon of the ingroup branches off next. How can you decide which of the ingroup taxa should branch off next? Look at your matrix and you will see that three of the four ingroup taxa have four characters in their columns while Plantae has five. You now have to make a judgment call. Since this may be your first try at cladistics call it a hypothesis and choose Animalia, Fungi or Protista to branch off next. Feel free to use prior knowledge and common sense as you do this – that's what brains are for. For the purposes of this example, I decided to have Protista branch off next. Protista has three traits that Prokaryotes did not: Nucleus, Chloroplasts and Cell Walls.
e. Decide which of these characters should go on the backbone between Prokaryotes and Protista and which should go on the side branch to Protista. Nucleus is easy. It goes on the backbone because all remaining taxa have this trait. The other two traits are tougher because all remaining taxa do not have Chloroplasts and Cell Walls. You can handle this in different ways; two of these are shown on the cladograms in **Figures 1.1** and **1.2**.
f. Continue adding characters and branching off taxa until all taxa and characters are accounted for on the cladogram.
g. The cladograms in **Figures 1.1** and **1.2** were both produced using the same matrix. Check to make sure that all taxon and trait combinations are present on both cladograms. Which cladogram is correct? Is either correct? The first question you should ask is, "Are they equally parsimonious?" They both have eight character changes, i.e., places where characters appear or disappear so they are equally parsimonious. Even so the cladograms show fundamentally different evolutionary scenarios.
h. Take time with your partner/team to review both cladograms while referring to your matrix. Compare and contrast the two cladograms and think about why the characters were placed where they were. Discuss these differences between the cladograms. WRITE your observations in your lab notebook.

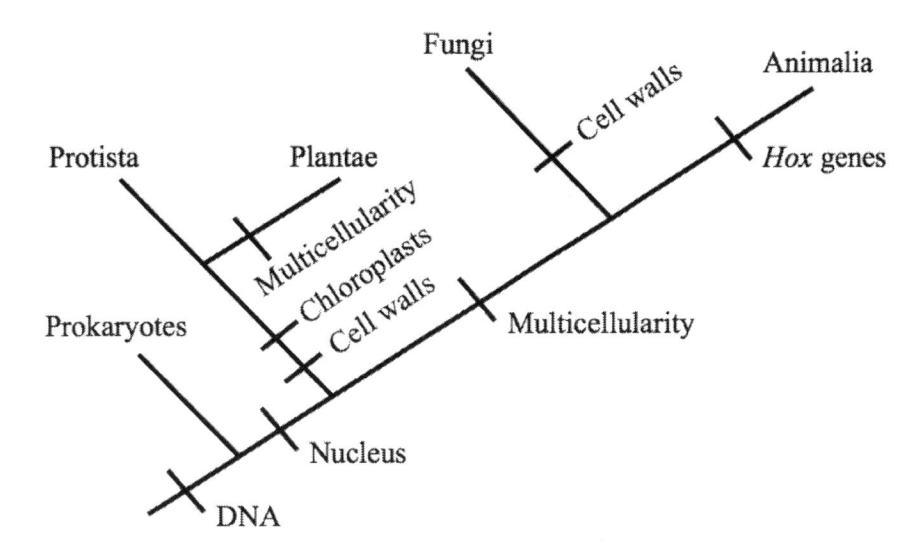

Figure 1.1. Cladogram #1, developed from the data provided above. (Image: ARH)

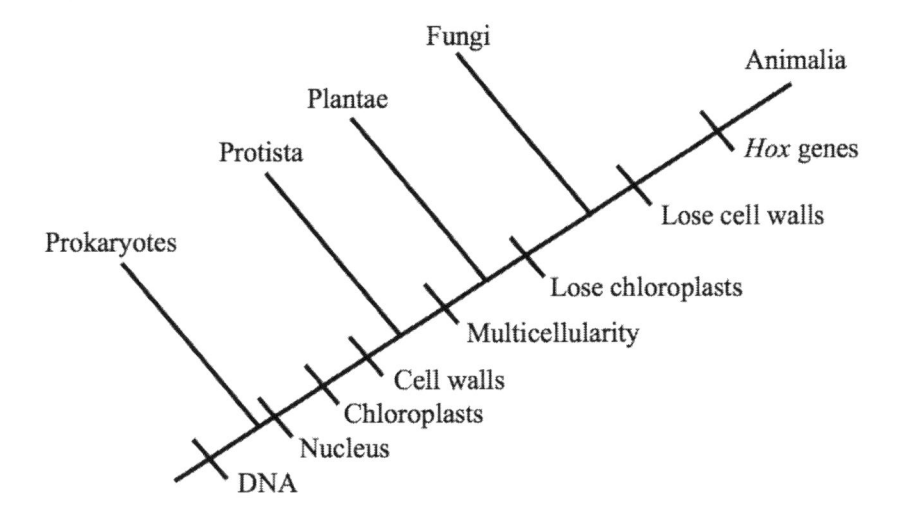

Figure 1.2. Cladogram #2, developed from the data provided above. (Image: ARH)

Task #3: Produce a cladogram

Now that you have been introduced to the terminology and basics of cladistics it's time for you to do some phylogenetic analysis.

1) Work as a member of a team to carry out this exercise. Your instructor will let you know how many students will be in each team.
2) Your instructor will inform you which animals from those in **Fig. 1.3** will be included in your analysis. Note that the jellyfish has been designated as the outgroup.

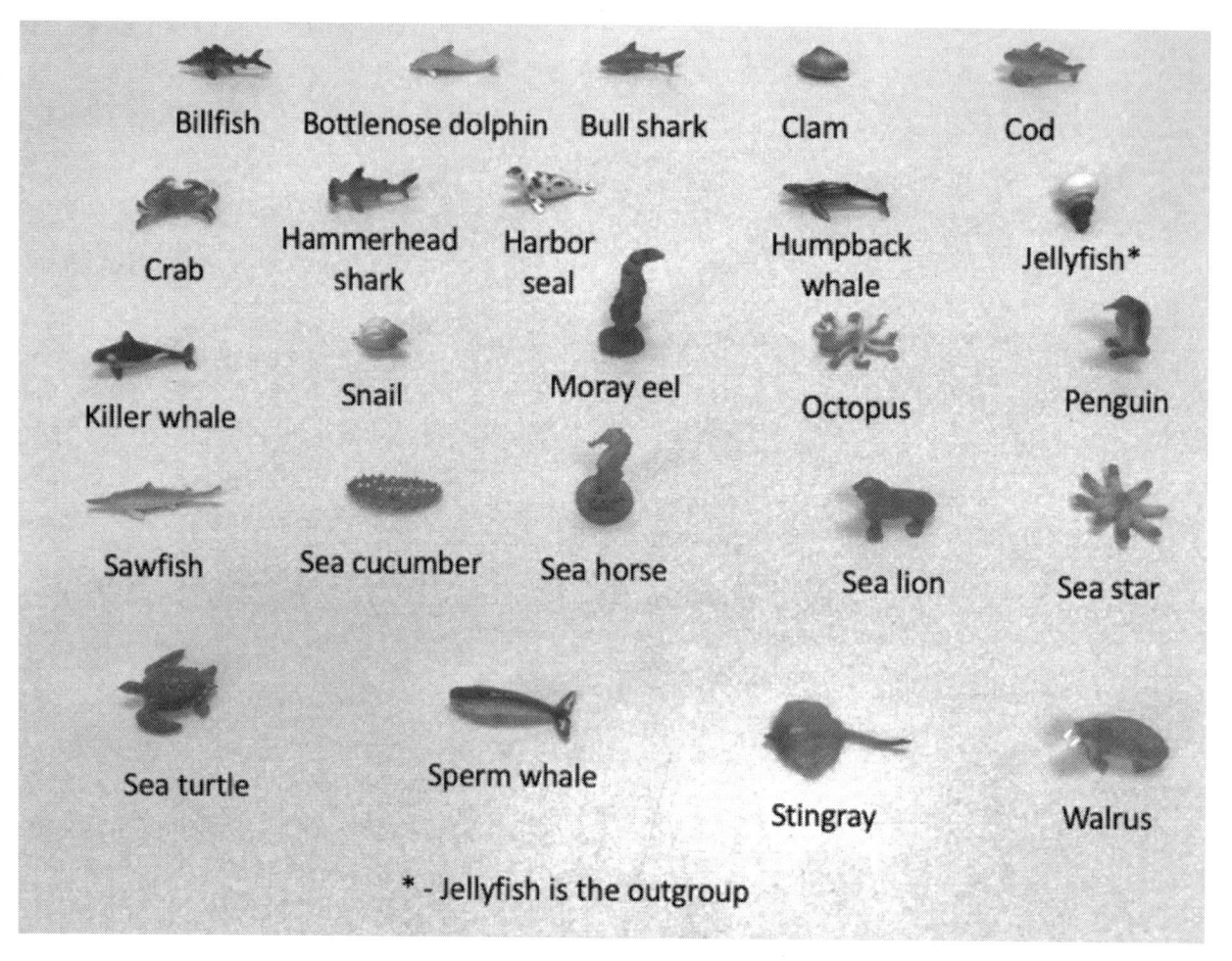

Figure 1.3. Animals and their common names that may be used to produce the cladogram. The asterisk indicates that the jellyfish is the outgroup. (Image: ARH)

3) Identify a characteristic that all animals share and use it to define the clade that includes all animals in the analysis. Do not use behavioral traits in cladistics: all traits for this exercise must be anatomical. Feel free to consult other sources for characteristics of each animal as you work.

4) Identify another characteristic that is found in all animals except the jellyfish and use it to define a clade that includes all animals except the jellyfish. Feel free to arrange and re-arrange the plastic animals into different groups based on synapomorphies. This exercise, as in all cladistic work, is a process of hypothesis testing.

5) Identify characteristic after characteristic to designate progressively smaller clades until all animals in the set have been accounted for. Remember that a side-branch may bear more than one taxon. But each animal in the analysis must ultimately have at least one synapomorphy that differentiates it from all other animals in the analysis. Don't forget that any trait placed on the backbone of the tree applies to everything else up the tree. Also, a trait may be gained and lost, but not lost and regained.

6) Use the characters you identify and taxa to develop your tree and associated matrix. All taxa and characters must be included in the tree. The overall goal is to produce a tree that

has the fewest number of character-changes possible. The smallest number of character changes possible in any tree is n-1 changes where n=the number of taxa in the analysis, though n-1 character changes is not always possible to achieve.

7) For this exercise students often find it easier produce the tree and then generate the matrix, though this is not how most cladistics is done.
8) INCLUDE your final matrix and cladogram in your laboratory notebook along with observations and questions you have about generating cladograms.

The Laboratory Notebook

Your instructor will let you know if you are required to produce a laboratory notebook as part of your lab experience. And if so, s/he will provide you with the specific format they want you to follow. If you are not required to have a lab notebook as a course requirement, I strongly encourage you to keep one anyway. The information below includes directions about one way to format a laboratory notebook.

A laboratory notebook should contain a complete record of the work you do in lab (just like a research notebook should be a complete record of your research). The main goal of producing a lab notebook is to help you learn to work and think as a scientist, and to practice the skills needed to make and record scientific observations. The benefits of keeping accurate records and developing the skills needed to produce a well-formatted notebook cannot be overstated, especially if you plan to pursue a career in the sciences. The ability to do this will yield great dividends when you begin carrying out your own research and as you collect and record your own observations and data.

Choose a spiral-bound, hard-covered artist's sketchbook that is at least 7"x10" in size for your notebook (larger is better). The hard cover protects your work, and the spiral binding allows the notebook to lie flat on a table or desktop while you work. The paper in an artist's sketchbook is acid-free so it will not yellow or become brittle over time. You also need 2H or 3H pencils. The lead in regular #2 (HB) pencils is too soft and will smear over time, so don't use them. Pencils with 2H or 3H lead produce a nice line and will not smear easily. And NEVER use a pen or colored pencils or markers in your lab notebook.

You should develop your lab notebook as you work. DO NOT make your laboratory notebook a re-copy notebook where you do work on loose sheets of paper or in another notebook and then recopy your work into the notebook later in an effort to make it prettier. The notebook is designed to be a working tool that you develop throughout the entire course. It frankly doesn't matter how pretty your notebook is as long as it is complete, well organized, and contains a complete record of the work you did in lab.

Task #4: Formatting guidelines for the lab notebook

1. <u>Contact information</u>: Write your name and enough contact information inside of the front cover of your notebook so that if you happen to lose or misplace it, whomever finds it will have a way to return it to you. E.g., email address, phone number, etc.

2. <u>Microscope ocular micrometer calibrations or aperture field width measurements</u>: Write the ocular micrometer measurements of the kind of microscope you use inside the front

cover of your notebook. If microscopes you use don't have ocular micrometers, write the field width measurement for each magnification inside the front cover of your notebook. You will refer to these measurements as you generate a scale bar for every drawing you make. Having this information where you can find it quickly and easily will streamline the work you do.

3. <u>Table of Contents</u>: Leave the first two or three pages of the notebook blank when you make your first entry. These pages will be used for the Table of Contents. A well-designed Table of Contents is essential to quickly and easily finding entries in your notebook. Each entry in the Table of Contents should include a page title, page number and the date the work was done. You can include each page in the Table of Contents or you can include each week's lab title instead (the latter is preferred).

4. <u>Page formatting for the notebook</u>: Always write the page number, date and a page title at the top of *every* page of the notebook. A page title is just a brief description of what's on the page. It can be as simple as "Lab #1 Phylogenetic Analysis" or a specific description of whatever is on a particular page, e.g., "Phylogenetic Analysis Matrix".

5. <u>Page layout</u>: For pages containing a lab drawing, divide the page into three sections by drawing a line across the page 6 - 8 cm from the bottom of the page, and then draw a vertical line from this line down to the bottom of the page, dividing this area into two smaller panels. The large top panel will contain the drawing. One of the smaller panels at the bottom is for your observations, and the other smaller panel is for questions you generate in relation to the entry you make on this page. See guidelines for observations and questions later in this section.

Laboratory Drawings

You will produce many drawings during this course. Why? Most people roam the planet looking at lots of things but actually seeing very little. For example, the average person looks around and sees a house, a tree, a person, a dog, etc., but they usually don't go any farther than that. They don't look for details. Artists, on the other hand, take this one step farther; they look for additional information. They look at a tree and they see patterns of light and dark, the texture of the bark, the angles and patterns of branching, the sizes and shapes of leaves, branches, etc. That is, artists see what they look at. Scientists take this another step beyond that and look for enough information to help them try to explain what they see. In other words, scientists strive to see, describe and explain what they see. They do this by making observations and asking questions. For example, a scientist may ask, "Is the branching pattern of this tree adaptive, and if so, how?" Scientists then collect observations and use their data to try to answer their questions. Producing laboratory drawings helps you develop and hone the observational skills of an artist and the curiosity of a scientist. Your work pattern during lab should be to look, see, draw and then ask questions. As you engage in this pattern of observing, drawing and asking, you will develop observational and question-asking skills used by scientists.

To answer the earlier question about why you are required to draw, drawing forces you to look for detail. It has been said that a pencil is a great aid to observation. Most of the drawings you will produce are assigned in the laboratory exercises, but in order to gain full benefit from the lab experience you should push yourself to produce additional drawings of what you see. Remember, the goal of producing lab drawings is to help you observe what you are looking at, as well as to provide a record of what you did during lab.

Lab drawings are working drawings and as such need not be artistically beautiful, though they should include as much detail as possible. Drawings also help you recall what you saw and did in lab. To that end lab drawings should represent what *you* saw, not what is shown in drawings or photographs in the lab exercises or in the textbook. Draw what you see. *Never* replicate a drawing from the lab manual to complete an exercise. The entries in your lab notebook should provide you with a good enough record of your experiences so that if the occasion requires, you can turn to your lab notebook for review rather than going back to the actual specimens in lab.

Task #5: Guidelines for laboratory drawings

1. Draw big! Fill at least half of the page with one drawing, and unless otherwise directed you should have only one drawing per page. This is important because the larger a drawing is, the more detail you can include in it. Drawing large also forces your eye to look for detail you might otherwise overlook. Include everything you see in your drawings even though you may not know what everything is. This will in turn lead you to asking better questions.
2. Produce a scale bar for each drawing.
3. Whenever possible, include more than one drawing from a single specimen/dissection.
4. Always use a 2H or 3H pencil. Never use a pen or marker for anything in your notebook.
5. Do not render (i.e., shade or color) your drawings. Rendering often obscures detail rather than enhances it. Instead, describe the things you see via accompanying observations.
6. Some students are tempted to use a cell phone to photograph specimens and refer to these later to make their drawings, but photographs cannot replace good first-hand observations. Plus, they often lack depth of field needed to see everything.

Task #6: Guidelines for Recording Observations and Questions

Label all the structures you can on each drawing you make. You are also required to generate observations and questions for each drawing/entry you make in your notebook. Observations can clarify what you see as you develop a drawing or they can be thoughts you have about your specimen as you study it. Questions can be about any aspect of the specimen you are observing. They will, I hope, be insightful and often lead you to additional questions and observations.

If you are not sure where to start when it comes to posing questions, consider questions that start with "I wonder…(how/if/what/why/when)". Also, any question you would ask your neighbor or the instructor about your specimen could and perhaps should be jotted down.

As a rule of thumb, you should have at least three observations and three questions for every drawing in your laboratory notebook.

Group Questions

Discuss and answer these questions in groups of two to four students. Write your answers to these questions in your lab notebook.

1) In what ways are the cladograms in Figures 1.1 and 1.2 alike? In what ways are they different?
2) Reflect on the cladograms in Figures 1.1 and 1.2 and describe the different evolutionary scenarios (stories) they show.
3) Comment on any unexpected results that appeared in your team's cladogram.
4) Explain why phylogenetic analysis produces large clades that include progressively smaller clades that are nested within each other, analogous to a stacking doll (see below).

(**Image**: Modified by ARH from an image by Fanghong under GFDL license - Creative Commons Attribution-Share Alike 3.0)

Chapter 2: Introduction to Microscopy and Plant and Animal Cells

In this chapter, you will be introduced to principles of microscopy, and you will have opportunities to examine plant and animal cells. Many of you already have some experience using microscopes, but it is important that you receive adequate instruction in microscopy so you can take full advantage of these important tools and then pass these skills along to others as opportunities arise.

Introduction to Microscopy

About now you might be thinking, "Why do *I* need an introduction to microscopy? I already know how to use a microscope." I don't doubt that many of you have had chances to use microscopes, but sadly, the instruction many students receive about microscopy is incomplete or inadequate. The purpose of this section of this chapter is to provide you with more complete instruction in microscopy than you may already have had, and to help you learn practices and skills used by good microscopists.

Task #1: Become familiar with a compound microscope

Compound microscopes are used to examine specimens that are small enough, thin enough or transparent enough to allow light to pass through them. Compound microscopes have a light source in the base of the microscope that projects light up through a specimen and into the lens system. Use **Fig. 2.1** and the video link below to become familiar with the parts of a compound microscope: https://www.youtube.com/watch?v=atj5vZPgcl8

Parts of the compound microscope

    Arm – This structure supports the head of the microscope.

    Base – The base supports the microscope and houses the lamp.

    Coarse/fine focus knobs – The larger outer knob is the coarse focus. Use this knob to find the focal plane for each magnification. The smaller inner knob is the fine focus. Use this knob to fine-tune the focus once you have found the focal plane.

    Condenser – This is located beneath the stage; the condenser contains a lens that focuses light from the lamp to improve image quality.

    Condenser adjustment knob – This knob moves the condenser up and down.

    Head – The head of the microscope houses a set of mirrors that reflects the light from the objective lens into the ocular lens(es). This lens system causes the specimen to appear upside down and backwards. The microscope head bears the ocular lens and a rotating set of objective lenses. You should take note of the magnification power of each of these lenses. If you loosen the thumbscrew to adjust the orientation of the head, make sure that you re-tighten it before you start using the microscope.

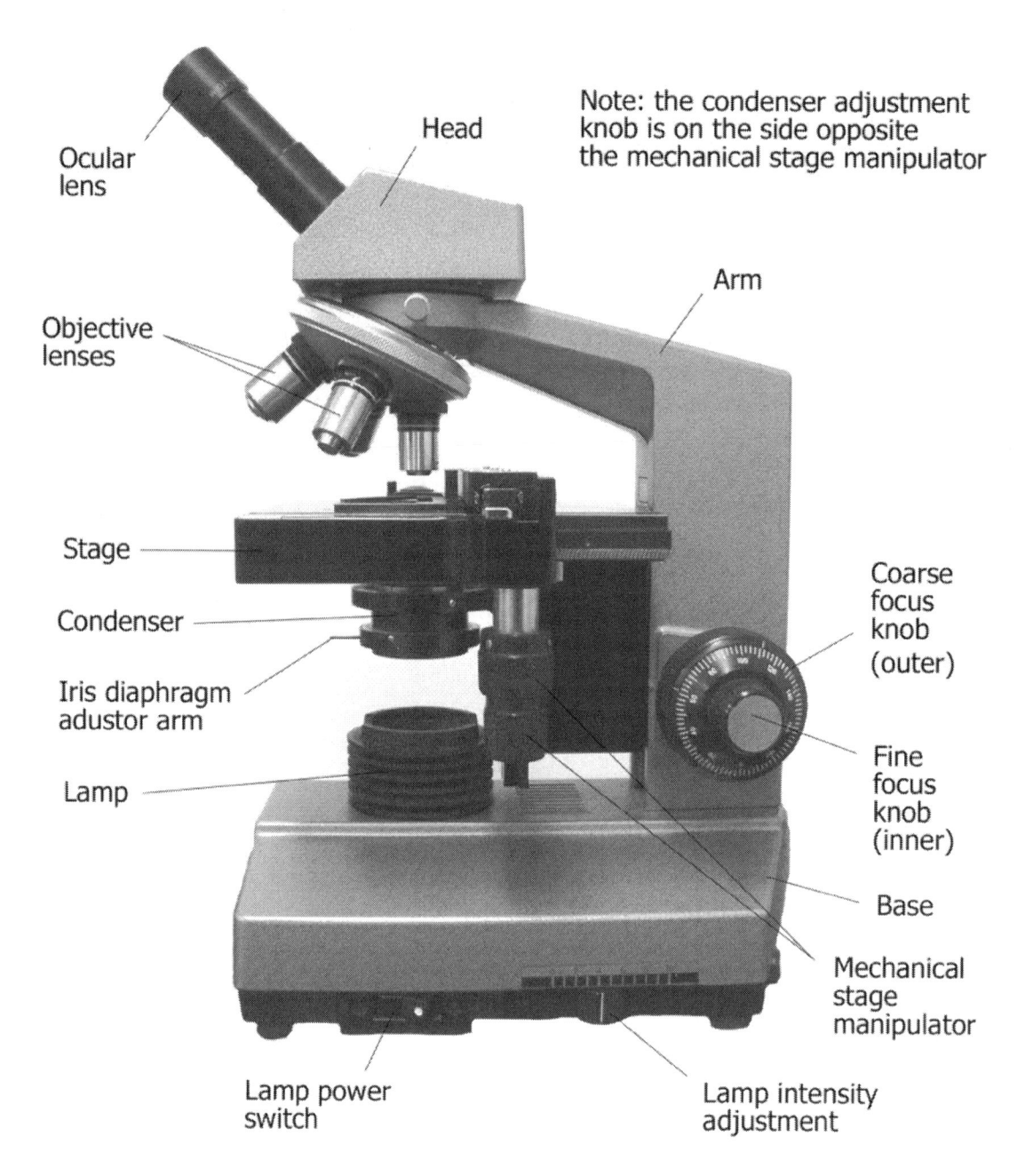

Figure 2.1. A monocular compound microscope. (Image: ARH)

Iris diaphragm and adjustor arm – The iris diaphragm is used to regulate the amount of light passing from the lamp through the condenser to the objective lens. The adjustor arm dilates and constricts the iris diaphragm opening. Image contrast increases as the iris diaphragm is closed.

Lamp – Light from the lamp passes up through the specimen and into the lens system. The notch in the lamp housing is used when you adjust the condenser.

Lamp intensity adjustment – The intensity of light produced by the lamp is adjusted by a sliding lever or knob, depending on the kind of microscope you are using. Increase lamp intensity only enough to see the image clearly. If the light is so bright that you feel pain or discomfort while looking through the ocular lens(es), decrease the lamp intensity.

Lamp power switch – This switch turns the lamp on and off. This switch should always be turned to the off position when you are done using the microscope.

Mechanical stage manipulator – These knobs move the spring-loaded slide holder forward and backward and side-to-side across the stage. Use your finger to pull back the spring-loaded arm of the slide holder, insert a slide, and then gently release the arm. This will hold the slide in place as you observe it.

Ocular lens or eyepiece – The ocular lens(es) slide into the barrel(s) on the top of the microscope head. The magnification of the ocular lens is indicated on the lens and is typically 10x, but sometimes 15x. This magnification should be noted so you can calculate overall magnification while you are using the microscope. If you are using a binocular compound microscope, one of the lenses is often fixed (can't be adjusted) and the other may be focused independently. If this is the case for your microscope, you will need to focus on the specimen on the slide stage using the fixed focus lens (the one you can't adjust) first, and then use the adjustable focus of the other lens to achieve the best focus for both lenses.

Objective lenses – This set of lenses is mounted on a rotating base located on the underside of the microscope head. The magnification power is inscribed on each lens. You should take particular notice of the magnification power of each lens, and of any lens or lenses that are indicated as oil-immersion lenses. Usually, there is only one of these, and is typically the 100x objective lens. Oil-immersion lenses are the only lenses that should be used with immersion oil. If immersion oil comes in contact with any other lenses, it will ruin them. Your lab instructor should demonstrate how to use immersion oil.

Stage (with spring-loaded slide holder) – The stage supports the slide, and has a hole in the center where light from the lamp passes up toward the objective lenses.

Task #2: Become familiar with a dissection microscope.

Dissection microscopes are used for specimens that are either too large or too opaque to observe using a compound microscope. When using a dissection microscope, light reflects off of the surface a specimen and up into the microscope, though some may also have a lamp in the base. Use **Fig. 2.2** and the video link below to become familiar with the structure and function of a dissection microscope: https://www.youtube.com/watch?v=mWdRyXVH5Cg

Parts of the dissection microscope

Focus knob – This knob is used to adjust focus.

Head, Arm, and Base – These parts fill the same functions as those on the compound microscope, though some dissection microscopes use external light sources rather than a light in the base.

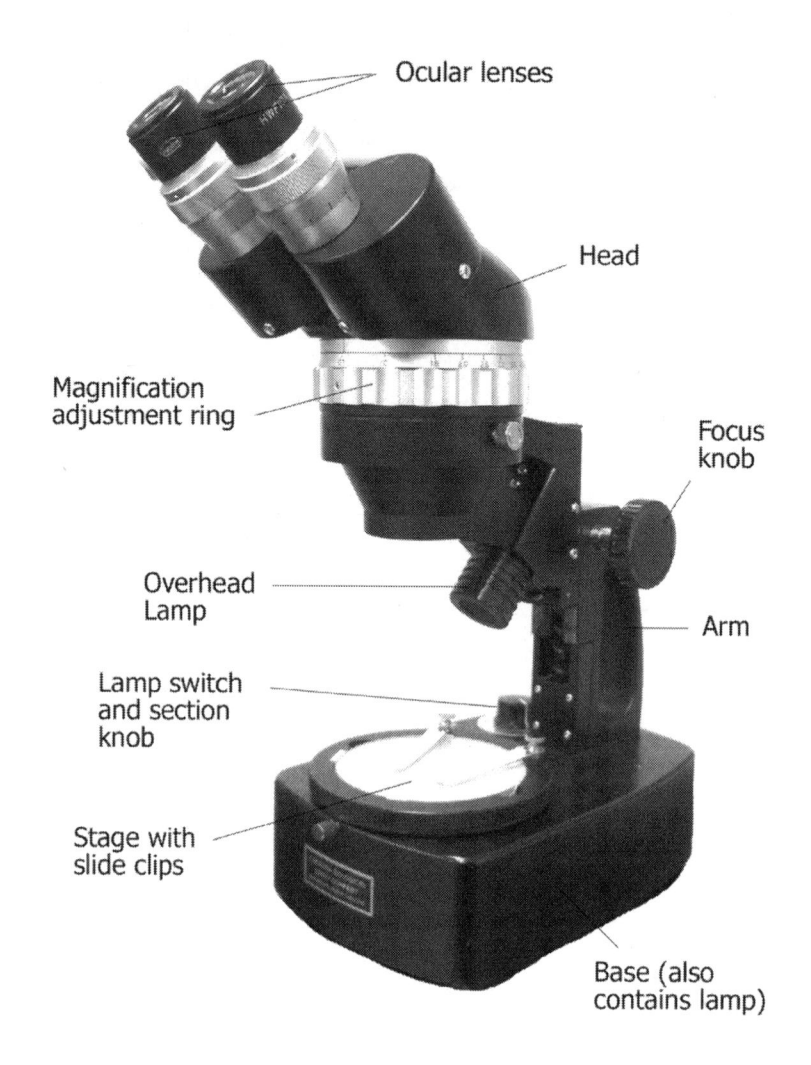

Figure 2.2. A binocular dissection microscope. (Image: ARH)

Lamp switch selection knob – This switch allows the user to choose whether to use a built-in overhead lamp, a lamp in the base, both, or neither (if present on the scope). Light intensity may not be adjustable on your dissection microscope.

Magnification adjustment ring – This rotatable ring on the microscope in **Fig.2.2** allows the user to adjust the zoom magnification between 0.7x to 4.5x (multiplied by the power of the ocular lens). On other dissection microscopes, magnification may be changed by rotating a knob located on the side of the microscope head.

Ocular lenses/eyepieces – These have the same function as on the compound scope. The ocular lenses of a binocular scope can usually be focused independently. Use one eyepiece to focus on a specimen, and then focus the other eyepiece as needed.

Overhead lamp – The lamp projects light down onto the specimen. Some dissection scopes do not have a built-in overhead lamp. Instead, you may have a stand-alone external lamp with flexible arms so you can direct the light on the specimen as desired.

Stage with stage clips – The stage supports the specimen during observation. Stage clips hold slides in place. Stage clips can be easily removed. The centerpiece of the stage can be opaque white, opaque black, translucent, or clear glass, and can be changed as needed.

Task #3: Basic Rules of Microscopy

This section will introduce you to basic rules of microscopy that will help you transport and use microscopes safely. These skills are hallmarks of good microscopists. Use the information below as well as this video to learn these rules:
https://www.youtube.com/watch?v=eJeEjS2aYPo

1) <u>Handling microscopes.</u> Always use two hands whenever you carry a microscope: one hand goes under the base to support the microscope and use the other hand to firmly grip the arm of the scope. Cradle the microscope in front of you as you carry it. This keeps the microscope away from table corners, doorknobs, etc., and makes it easy for others to see that you are carrying a microscope so they can clear a path for you.

2) <u>Cleaning lenses.</u> Clean ocular and objective lenses *only* when they need it. When a lens is dirty use *only* lens paper to clean it – never use any other kind of tissue, paper or cloth to attempt to clean a lens because these can scratch and permanently damage lenses. Lens paper is specifically designed to clean a lens without damaging it.

3) <u>Examining a specimen using a compound microscope.</u> To examine a specimen, move the slide stage to its lowest position and place the slide on the stage. Rotate the lowest power objective lens into position and then move the stage to its highest position. While looking through the ocular lens, use the coarse focus knob to move the slide stage down and away from the objective lens and locate the plane of focus. Moving the slide away from the objective lens prevents you from moving a slide up into an objective lens, which could break a slide and possibly damage the lens.

4) <u>Finding the fine focus.</u> Once you have found the focal plane using the coarse focus knob, use the fine focus knob to fine-tune the focus. You should never rotate the fine focus knob more than one full rotation in either direction to achieve fine focus on your specimen. If you need to rotate the fine focus knob more than one full turn in either direction to adjust focus, start over with the coarse focus knob.

5) <u>Increasing magnification.</u> Always work your way up through the objective lenses, in order, from lowest power to the desired power as you examine a specimen. If you skip an objective lens, you may have a difficult time finding the plane of focus again. Compound microscopes are built so that as you rotate one objective lens out of position and another higher-powered lens into position, you should be very close to the same focal plane. If you are using the dissection scope, start with the lowest magnification possible, and increase magnification as needed.

6) <u>Using immersion oil.</u> Use immersion oil only if you need to use the immersion oil (100x) objective lens. Oil immersion lenses will ALWAYS be labeled as being immersion lenses and/or have a black line around the tip of the lens. If you have any

questions about this, *ask your lab instructor before you do anything.* Immersion oil is specifically designed to be used only with immersion oil lenses, and this oil will damage other lenses. Immersion oil helps produce better images at the highest magnification because once you reach this power of magnification, the field of view is so small that only a small amount of light can enter the objective lens due to light scattering. Immersion oil collects light and directs into the objective lens with minimal scattering.

Follow this procedure while using immersion oil:
1. Work your way up through all objective lenses until you find the focal plane for the 100x lens.
2. Rotate the objective lenses so the lowest power objective lens and 100x lens are on either side of the slide being observed.
3. Use the built-in applicator in the jar of immersion oil to apply one or two drops of immersion oil to the top of the coverslip of the microscope slide (you should let excess oil drip off of the applicator and back into the bottle before moving it over the slide)
4. Once there is a droplet of oil on the slide, rotate the 100x lens into that drop – WARNING – always double-check to make sure that you are rotating the 100x lens into the oil.
5. If you lose the plane of focus during this process, rotate the 100x lens out of position, remove and clean the slide and the objective lens, and start over.

7) Preparing a microscope for storage after use. Do the following every time you are done using a microscope.
   1. Remove the slide or specimen from the stage.
   2. Move the stage to its lowest position.
   3. Clean the oil immersion lens (only if you used it, and use only lens paper)
   4. Rotate the lowest power objective lens into position
   5. Turn the lamp off.
   6. Wrap the cord loosely around the base.
   7. Replace the dust cover (if one is provided).
   8. Using two hands, return your microscope to its storage location.

Task #4: Setting up a compound scope for use (do this every time before using a microscope)

1) Check the microscope:
   1. Plug the scope in and make sure the lamp works
   2. Move the stage to its lowest position
   3. Make sure a slide was not left on the stage by the previous user.
   4. Rotate the lowest power objective lens into place.
   5. Check the lenses to make sure they are clean; clean them ***only*** if needed (you should only rarely need to clean any lenses).

2) Adjust the condenser:
   1. Use the lowest power objective lens to focus on a prepared slide.

2. Use the condenser adjustment knob to move the condenser to its highest position
3. Place the tip of a probe so that it rests horizontally through the notch on the lamp housing so that its tip touches lightly but directly on the center of the lamp.
4. Look through the ocular lens while holding the probe tip in position and, using the condenser adjustment knob, rotate the condenser downward until you see a sharp outline of the silhouette of the probe tip in the field of view.
5. When the silhouette of the probe tip is in sharp focus you will probably also see a mottled pattern in the rest of the field of view. This mottling is produced by imperfections in the glass surface of the lamp. Use the condenser adjustment knob to move the condenser either slightly up or down, just enough so that the mottling disappears. Your condenser is now correctly adjusted and your microscope is ready to be used.

3) <u>Calibrate the ocular micrometer</u>: Your instructor will let you know if you need to do this – and you need to do this only once, not every time you use a microscope

An ocular micrometer is a small glass disc located in the barrel housing one of the ocular lenses. You need to calibrate the ocular micrometer only once if you are using the same microscope every time, or if you are using the same kind of microscope, like in a lab teaching set. During calibration you need to determine the actual distance between the vertical lines on the ocular micrometer for each magnification (each objective lens) on your microscope. To do this you will need a slide called a stage micrometer. The markings on an ocular micrometer will look like those shown in **Fig. 2.3**. The markings on a stage micrometer look like those shown in **Fig. 2.4**. You should note that the divisions on the ocular micrometer are numbered, but there is no indication of the distance between them. This is because the distance between vertical markings changes with each magnification, since the area in the field of view decreases as magnification increases.

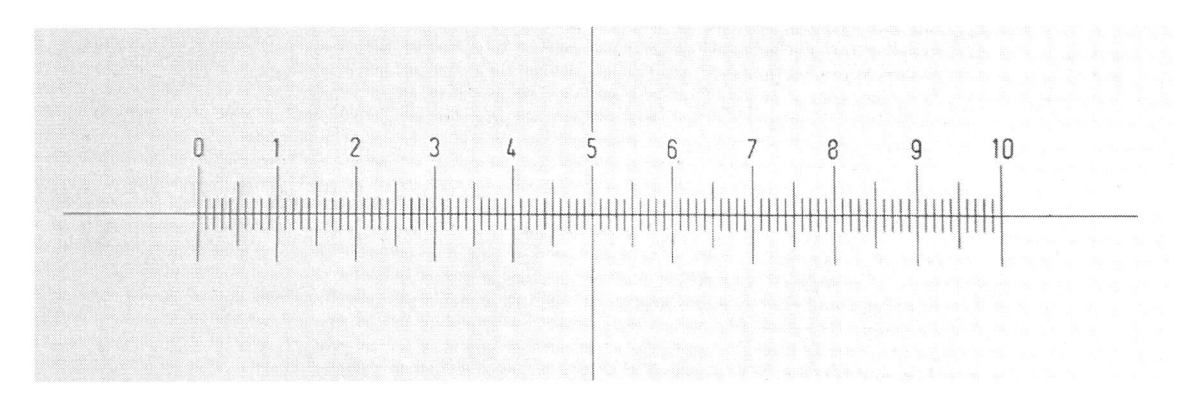

Figure 2.3. Ocular micrometer, divided into 100 divisions called ocular micrometer units = o.m.u.

The goal of calibrating the ocular micrometer is to determine the width of an ocular micrometer unit (o.m.u.) for each objective lens so you can use it to measure the sizes of things at each magnification. Here's how to do this:

1. Put the lowest power objective lens in place and focus on the stage micrometer. The stage micrometer shown in **Fig. 2.4** is 1000 μm (1.0 mm) long, divided into 100 divisions of 10 μm (0.01 mm) each.

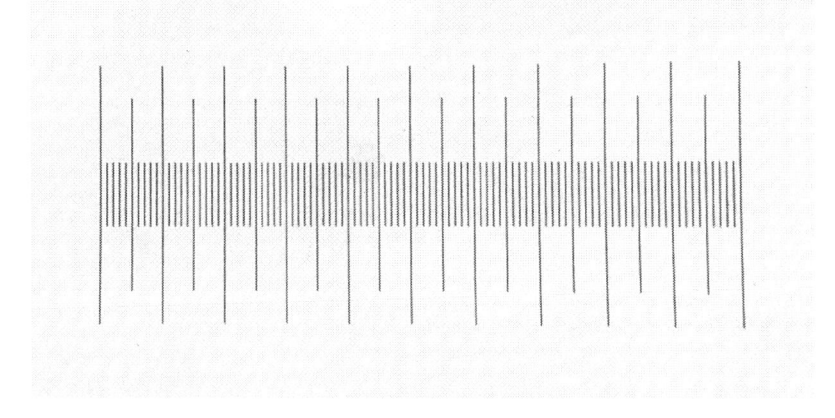

Figure 2.4. A 1000 μm (1.0 mm) stage micrometer divided into 100 ocular micrometer units of 10 μm (0.01 mm) each.

2. Rotate the eyepiece that houses the ocular micrometer so that the lines on the ocular micrometer and those on the stage micrometer are parallel to and partially overlap each other each other as shown in **Fig. 2.5**, with the far left-hand hashmarks of the stage micrometer and of the ocular micrometer aligned directly on top of each other.
3. Now look for two lines that line up perfectly with each other that are farthest from the left-hand end of the two micrometers. Count the number of divisions on the stage micrometer between the far left-hand end and the mark where the two hashmarks line up, and multiply that number by 10 μm.
4. Next, count the number of divisions on the ocular micrometer between the far left-hand end and the right-hand end where the marks line up.
5. Divide the calculated distance from the stage micrometer by the number of ocular micrometer units. This number is the physical distance for each ocular micrometer unit (distance between each set of o.m.u. hashmarks) *for this magnification.*
    - Example, using **Fig. 2.5**. The number 2 (the 20th hashmark) on the ocular micrometer lines up perfectly with the 50th stage micrometer hashmark.
    - Multiply the number of stage micrometer units by 10 μm = 500 μm
    - Divide the total distance from the stage micrometer (500 μm) by the number of hashmarks on the ocular micrometer (20). 500 μm/20 o.m.u. = 25 μm/o.m.u. at 4x magnification.

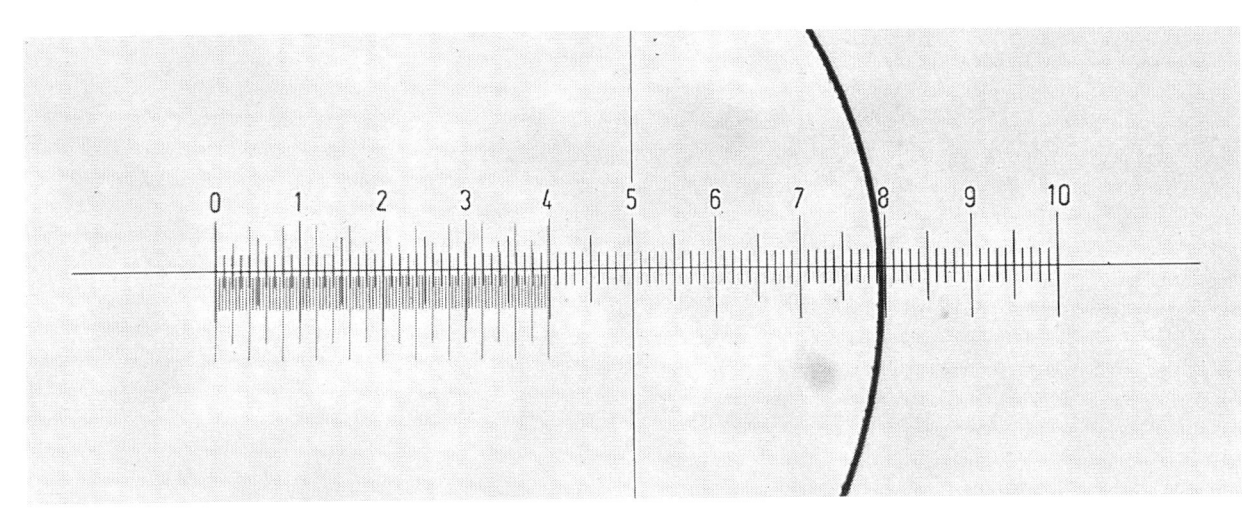

Figure 2.5. Ocular micrometer (with numbered major hashmarks) lined up with the stage micrometer using a 4x objective lens. Note that the number 2 on the ocular micrometer (20 o.m.u. from the left) lines up with the 5th large hashmark of the stage micrometer unit line (50 stage micrometer units x 10 μm each = 500 μm). So, using the 4x objective lens on this microscope, each ocular micrometer unit is 500 μm / 20 o.m.u. = 25 μm / o.m.u.

6. Repeat this procedure for each objective lens on your microscope (**Fig. 2.6** for a 10x lens and **Fig. 2.7** for a 40x lens; an image for a 100x lens is not shown). Once you have done this each objective lens on your microscope, write down the o.m.u. width for each magnification.
7. When you are finished calibrating your ocular micrometer for each magnification, return the stage micrometer and you are ready to use the ocular micrometer to measure specimens using your compound microscope.

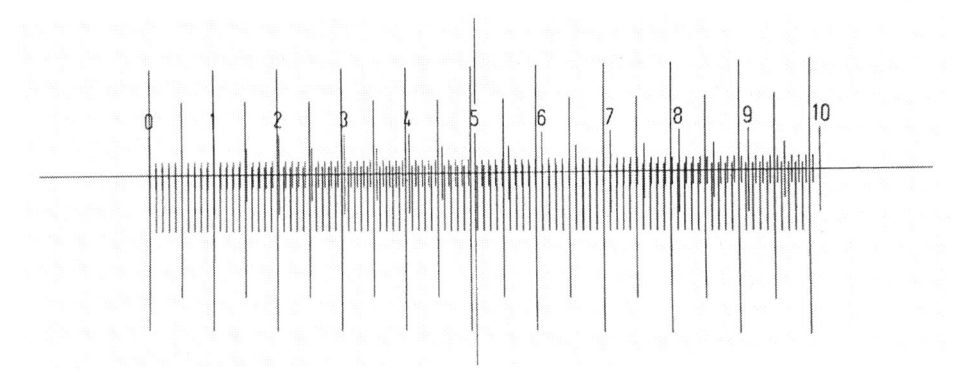

Figure 2.6. Ocular micrometer calibration using a 10x objective lens. The 69 o.m.u. hashmark lines up with the 70th stage micrometer unit hashmark. So, in this case, 700 μm / 69 o.m.u. = 10.1 μm / o.m.u. for the 10x objective lens.

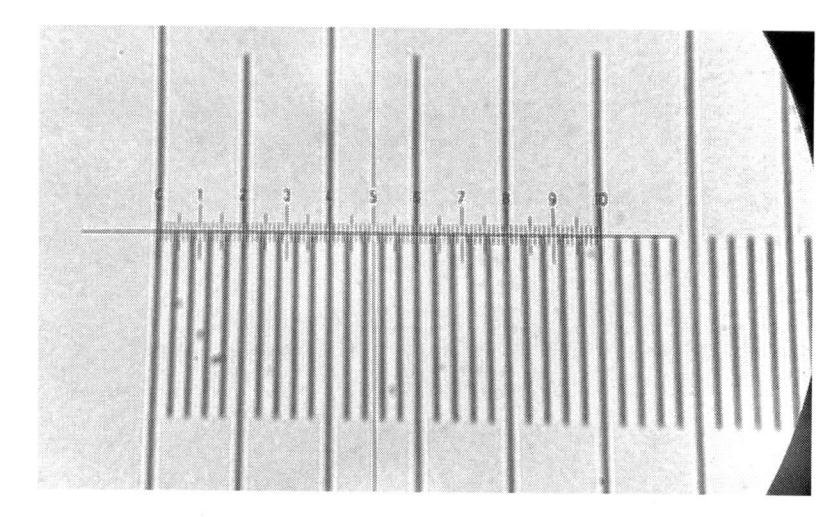

Figure 2.7. Ocular micrometer calibration using a 40x objective lens. The 60 o.m.u. line lines up with the 15th stage micrometer line (150 μm). So, for this microscope, while using the 40x objective lens, each o.m.u. represents 150 μm / 60 o.m.u. = 2.5 μm /o.m.u.

Task #5: Make a wet-mount slide

You occasionally need to make your own temporary slides of specimens. To make a temporary wet-mount slide you will need a specimen, a microscope slide, a cover slip, a pipette or eyedropper and some plasticene clay. Make a wet-mount slide by following these directions:

1) Place your specimen on the center of a glass slide. Then, using a pipette place a couple of drops of water on top of the specimen

2) Prepare the cover slip by holding some plasticene clay in one hand and a cover slip in the other. Gently drag each corner of the cover slip across the surface of the clay. A small bit of clay should now be on each corner of the cover slip. These small bits of clay serve as support posts to keep the weight of the cover slip from flattening the specimen.

3) Lay one edge of the cover slip down so it extends from one side of the microscope slide to the other with the plasticene clay on the underside of the coverslip.

4) Support the opposite edge of the coverslip with a probe tip. Lower the coverslip slowly until it comes in contact with the drop of water covering the specimen (see **Fig. 2.8**). Capillary action will draw the water out between the coverslip and the glass slide as you lower the coverslip. Lowering the cover slip slowly minimizes the number of air bubbles that get trapped under the coverslip.

5) If the cover slip floats there is too much water. Touch the corner of a paper towel to the space between an edge of the coverslip and the glass slide. The coverslip will settle onto the glass slide as water wicks out.

6) If water does not reach the edges of the coverslip, use a pipette to add more water to the space between the coverslip and the slide.

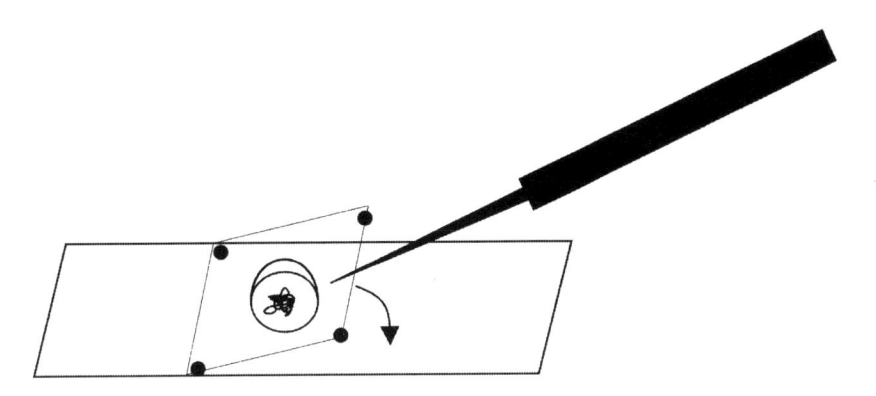

Figure 2.8. A wet-mount slide. The coverslip is lowered slowly until it comes in contact with the drop of water and then continues to be lowered until it rests on the microscope slide. Dots on the corners of the coverslip represent tiny amounts of plasticene clay (for purposes of illustration, these are much larger than they should be).

7) Your slide is now ready for observation. Keep in mind that the coverslip is not physically attached to the glass slide, so you need to handle the slide gently.

Task #6: Generate a scale bar for a drawing

A scale bar is used to indicate the scale of a drawing relative to the actual size of the specimen being observed. You should produce a scale bar for each drawing you produce in lab.

How to generate a scale bar for a drawing based on a macroscopic specimen:

1) Produce a drawing of the specimen you are observing
2) Measure some dimension, preferably the maximum length, of your specimen
3) Measure the same dimension on your drawing
4) Divide the length of your drawing by the length of the specimen. The resulting value is a number without units that is the scaling factor for your drawing.
5) Multiply this scaling factor by the length you want the scale bar to represent. Make it a nice round number such as 1.0 cm, 5.0 cm, etc., and the answer you get is the length of your scale bar for this drawing.

   Example:
   - The length of your drawing = 12.5 cm
   - The actual length of the object = 4.3 cm
   - The length you want the scale bar to represent = 1.0 cm
   - Scaling factor = (12.5 cm/ 4.3 cm) = 2.9
   - Scale bar length = 2.9 x 1.0 cm = 2.9 cm

6) Next, draw a line 2.9 cm long next to your drawing and then write "1.0 cm" above the line. This means that 2.9 cm on the drawing represents 1.0 cm on the original specimen, or, in other words, your drawing is nearly three times as large as the specimen.

How to generate a scale bar for a specimen observed using a compound microscope that has an ocular micrometer

1) Complete a drawing of the specimen you are observing under the microscope.
2) Measure some dimension on your drawing.
3) Use your microscope's ocular micrometer (that you previously calibrated) to measure the same dimension on the specimen you are drawing.
4) Divide the length of your drawing by the length of the specimen to generate the scaling factor.
5) Next, multiply the scaling factor by the length you want the scale bar to represent. Make the length to be represented a nice round number, like 100 μm, 10 μm, etc.

   Example:
   - The length of your drawing is 6.2 cm
   - The actual length of the object you drew is 448 μm
   - You want to produce a scale bar that represents 100 μm on the original object
   - Scaling factor = (6.2 cm / 448 μm) = 0.0138
   - Scale bar length = 0.0138 x 100 μm = 1.38 cm

6) Draw a line 1.38 cm long next to your drawing and write "100 μm" over the line. So, 1.38 cm on you drawing represents 100 μm on the specimen.

How to generate a scale bar if your microscope doesn't have an ocular micrometer. Note: This approach is fine for lab work, but it is not precise enough to do in a research setting.

1) Use a stage micrometer to measure the width of the field of view for each magnification (objective lens) of your compound microscope.
2) Draw a specimen you are observing under the microscope.
3) Measure the length of your drawing.
4) Estimate the length of the specimen you drew based on the field width. Note: This will not be especially precise, but you can get reasonably close.
5) To generate the scale bar for your drawing, do the same as you did above. That is, divide the length of your drawing by the length of the specimen to generate the scaling factor for your drawing.
6) Then, multiple the scaling factor by the length you want the scale bar to represent

   Example:
   - The length of your drawing is 8.7 cm
   - The actual length of the object you drew is 2575 μm
   - You want to produce a scale bar that represents 100 μm on the original object
   - Scaling factor = (8.7 cm / 2575 μm) = 0.00337
   - Scale bar length = 0.00337 x 100 μm = 0.34 cm

7) OOPS! For the example above, a scale bar 0.34 cm long is too short to draw with much precision, so in a case like this, increase the length that the scale bar represents to, say, 500 μm. Do this by multiplying the scaling factor by 500 μm instead of 100 μm. When you do this, you get 1.7 cm, a distance you can draw more precisely.

8) Draw a line 1.7 cm long next to your drawing and write "500 μm" over the line. This means that each 1.7 cm distance on your drawing represents 500 μm. Done!

Introduction to plant and animal cells

Task #7: Compare plant and animal cells
The fundamental unit of organization of every living thing is the cell. Internal and external anatomy of a cell varies with taxonomic group as well as cell function. In this exercise you will observe representative plant and animal cells.

1) Obtain either a prepared slide of animal cells provided by your instructor or make a wet mount side of cheek cells.
    a. Cheek cell preparation – make a wet-mount slide
        i. Use a tooth pick to *gently* scrape the inner surface of your cheek.
        ii. Swish the toothpick in a drop of water on a microscope slide.
        iii. Add the cover slip.
        iv. Add a drop of methylene blue dye to one edge of the cover slip and draw it across the specimen by touching a paper towel to the opposite edge of the cover slip.
    b. Observe, describe and DRAW the specimen available to you
    c. Generate a scale bar for your drawing.
    d. Label all parts of the cell you can observe. Refer to your textbook or online reference materials to assist you. It is likely that you will be able to see only the cell's nucleus and the cell's general shape.
    e. Write three observations and three questions in your lab manual related to each drawing you produce.

2) Obtain a plant cell specimen provided by your instructor. This may be a preserved specimen or fresh tissue preparation, such as onion epidermal cells.
    a. Onion epidermal cells
        i. Use a razor blade to remove a tiny piece of the thin onion bulb epidermis.
        ii. Make a wet mount slide of this tissue.
        iii. Add a drop of iodine to the edge of the cover slip and draw it across the specimen by touching a paper towel to the opposite side of the preparation, then add more water as needed.
    b. Observe, describe and DRAW your specimen.
    c. Generate a scale bar for your drawing.
    d. Include three observations and three questions about the specimen you observed. This is a standing assignment, so this reminder will not be included in other exercises.

Group Questions
1) What are advantages and limitations of using a compound microscope?
2) What are advantages and limitations of using a dissection microscope?
3) What is the benefit of producing a scale bar for each drawing you make?
4) Describe similarities and differences between plant and animal cells.

Chapter 3: Protists

Our classification scheme of biodiversity remained largely unchanged from the 1700s through the mid-1900s. This original taxonomy included only two kingdoms: Animalia and Plantae. In 1969 the Five-Kingdom Model of life was proposed, with Kingdoms Bacteria, Protista or Protoctista, Fungi, Plantae and Animalia. Then in 1977, the application of molecular methods to phylogenetic analysis led to the addition of the domains Bacteria, Archaebacteria and Eukarya at a taxonomic level higher than kingdoms.

Under this Three Domain Model, the eukaryotic kingdoms Animalia, Plantae and Fungi remain largely intact as monophyletic clades. However, Kingdom Protista was identified as being woefully polyphyletic and in dire need of review and revision. This is no surprise since Kingdom Protista has always been the eukaryotic garbage-can kingdom; the place where eukaryotic taxa were dumped that didn't fit easily into any of the other kingdoms.

Recent work on the taxonomy of Domain Eukarya resulted in a vastly improved taxonomy for all eukaryotes. Though this new taxonomy of eukaryotes is by no means final, a recently proposed taxonomic model (one of many that are out there) contains six eukaryotic super-groups: Amoebozoa, Archaeplastida, Chromaveolata, Rhizaria, Excavata and Opisthokonta (see **Table 3.1**). Taxonomic super-groups are placed between domains and kingdoms. One thing that can be said about the taxonomy of eukaryotes is that one should be surprised only if there are no further changes. Though Kingdom Protista is no longer accepted as a clade, the terms protist and protozoan are still used widely as terms of convenience.

Table 3.1. One proposed taxonomy of Domain Eukarya (*OpenStax Biology 2, 2018*).

Supergroup	Selected characteristics	Selected representatives
Amoebozoa	Mostly free-living unicellular and multicellular forms with broad, rounded pseudopodia, strictly heterotrophic	amoebae, cellular slime molds, plasmodial slime molds
Archaeplastida	Descendants of an endosymbiotic relationship between cyanobacteria and a heterotrophic cell.	red algae, green algae and plants
Chromalveolata	Descendants of a line that formed when a heterotrophic cell and red algal cell became symbiotic (lost in some).	dinoflagellates, diatoms, ciliates, brown algae
Rhizaria	Free-living, mixotrophic, parasitic and heterotrophic forms with long, thread-like filopodia, some secrete $CaCO_3$ or SiO_2 tests	foraminiferans and radiolarians
Excavata	Mostly heterotrophic forms with anterior flagellae, some mixotrophic and parasites, many with a ventral feeding groove	*Trichomonas, Giardia, Euglena, Trypanosoma*
Opisthokonta	Unicellular to multicellular flagellated forms with a single posterior flagellum that pushes the cell through the medium	fungi, choanoflagellates and animals

Task #1: Supergroup Amoebozoa (word roots = amoeba, animals)

Phylum Amoebozoa
1) Use a compound scope to observe, describe and DRAW the anatomy and behavior of any available live amoebae.
2) If living material is not available, describe and DRAW prepared specimens of *Amoeba*.
    a. *Amoeba proteus* (*Chaos diffluens*) - Free-living heterotrophic species in freshwater environments. Prey includes other protozoans, algae and micro-metazoans. Note the broad **pseudopodia**, also called **lobopodia**. The clear zone just within the plasma membrane and most easily visible at the tips of pseudopodia is the **ectoplasm**. Ectoplasm lacks organelles and other structures found in the **endoplasm**. Use **Figure 3.1** to help you identify what you see.

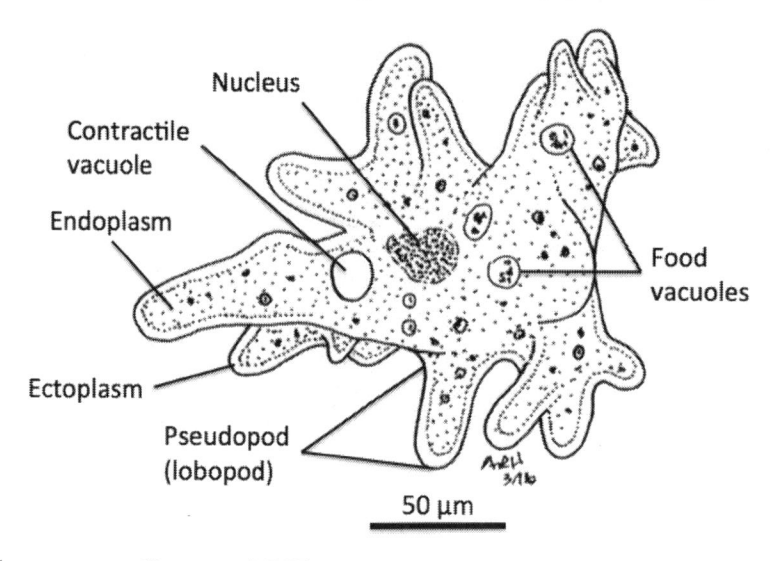

Figure 3.1. *Amoeba proteus.* (Image: ARH)

Task #2: Group Chromalveolata (word roots = color, pit)

Phylum Dinoflagellata (word roots = whirling, whip)

The dinoflagellates are a major component of both the marine and freshwater **phytoplankton**. They are unicellular flagellates that typically produce internal cellulose plates called **theca**. They have two flagellae that are perpendicular to each other: one produced in a transverse groove called the **cingulum** or girdle, and another in a longitudinal groove called the **sulcus**. This arrangement of flagellae causes a spinning motion of these cells as they swim. Thecae anterior to the cingulum are called the **epithecae**, and thecae posterior to the cingulum are called the **hypothecae**. Only about half of the species are photosynthetic, the rest being heterotrophic parasites or predators. Some autotrophic dinoflagellates are symbionts in the tissues of a variety of tropical marine animals, such as corals, where they form dense masses of cells and are called **zooxanthellae**. Dinoflagellates are typically most abundant in calm, nutrient-poor water. As nutrient levels and/or turbulence increases they are often replaced by other phytoplankton such as diatoms. Occasionally dinoflagellates form massive blooms called **red tides** that can cause extensive fish kills due to neurotoxins they release into the water. Some species are known for their bioluminescence.

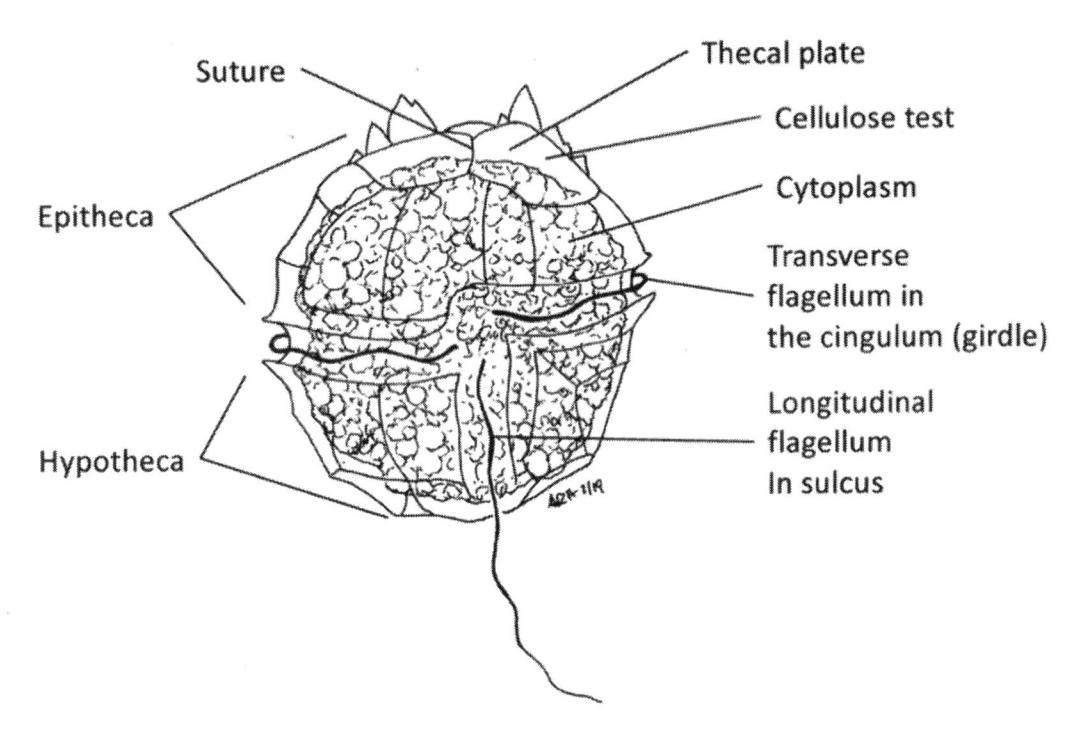

Figure 3.2. The dinoflagellate *Peridinium*. (Image by ARH modified from photograph by Picturepest, CC BY 2.0, https://creativecommons.org/licenses/by/2.0)

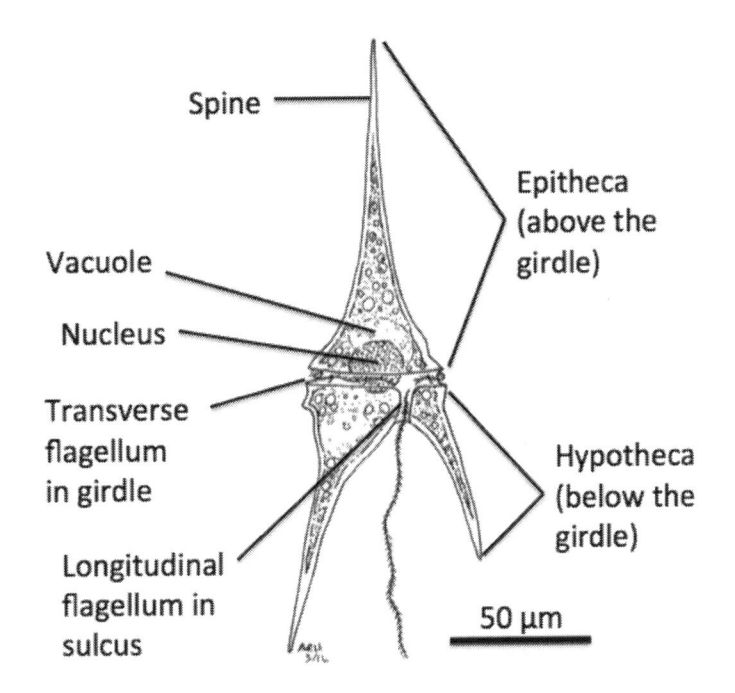

Figure 3.3. The dinoflagellate *Ceratium sp.* (Image: ARH)

1) Use a compound scope to observe, describe and DRAW any living dinoflagellate specimens. You may need to add Proto-Slow™ to your wet-mount slide to slow dinoflagellates enough to see them easily.
2) If living specimens are not available, use a compound microscope to examine prepared slides. Observe, describe and DRAW what you see. Two species that are often available are *Peridinium* and *Ceratium*. Use **Figures 3.2** and **3.3** to help you identify what you see.

Phylum Ciliata (word root = cilia, bearer)

1) Ciliates are protists that use cilia for locomotion, and are easy to identify at a glance because they glide smoothly through the water, not in a jerking motion like flagellates. Use a compound microscope to observe, describe and DRAW a wet-mount slide of ciliates from a live culture. Add Proto-Slow™ to your wet mount slides to slow these organisms down enough so you can observe them. Use **Figure 3.4** to help you identify what you see.
2) If live cultures are not available, study prepared slides. Observe, describe and DRAW what you see. One species you are likely to observe is *Paramecium*. Use **Figure 3.4** to help you identify what you see.

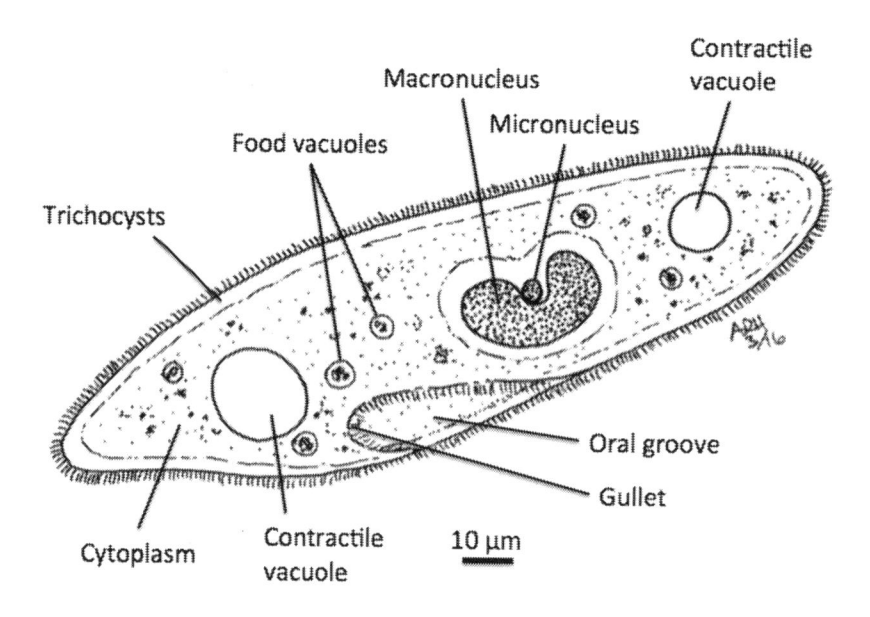

Figure 3.4 *Paramecium caudatum*. Cilia lining the **oral groove** pulls water, small cells and detritus into the gullet where these particles are put into food vacuoles. An individual may have many food vacuoles at once. This species has a **trichocysts** just below the outer surface of the plasma membrane. Trichocysts are believed to be mainly defensive in purpose and are harpoon or nail-shaped organelles that can be fired when *Paramecium* is threatened. (Image: ARH)

Phylum Bacillariophyta (word root = little stick, plant), Diatoms

Diatoms are the single greatest component of the marine phytoplankton and can be found in almost all aquatic environments where light is present. There are about 100,000 species of diatoms known and the number of undiscovered species could be many times that number.

Diatoms are unicellular protists that produce a silicon dioxide (SiO_2, i.e., glass) covering called a **frustule**. Each frustule consists of two halves or **valves** that look something like a petri dish with holes in it. There are two groups of Diatoms based on the symmetry of their frustules. Radially symmetrical species are called **centric diatoms** and those that are bilaterally symmetrical are called **pennate diatoms**. Diatoms usually store energy-rich oil in their cells, making them an important food source for many aquatic organisms. Most diatoms are planktonic, but many benthic species also exist.

3) Use a compound microscope to observe, describe and DRAW any representative diatoms available to you. Note: If you have live diatoms to observe, they will just look like brownish blobs – sorry. **Figure 3.5** shows examples of centric and pennate frustules.

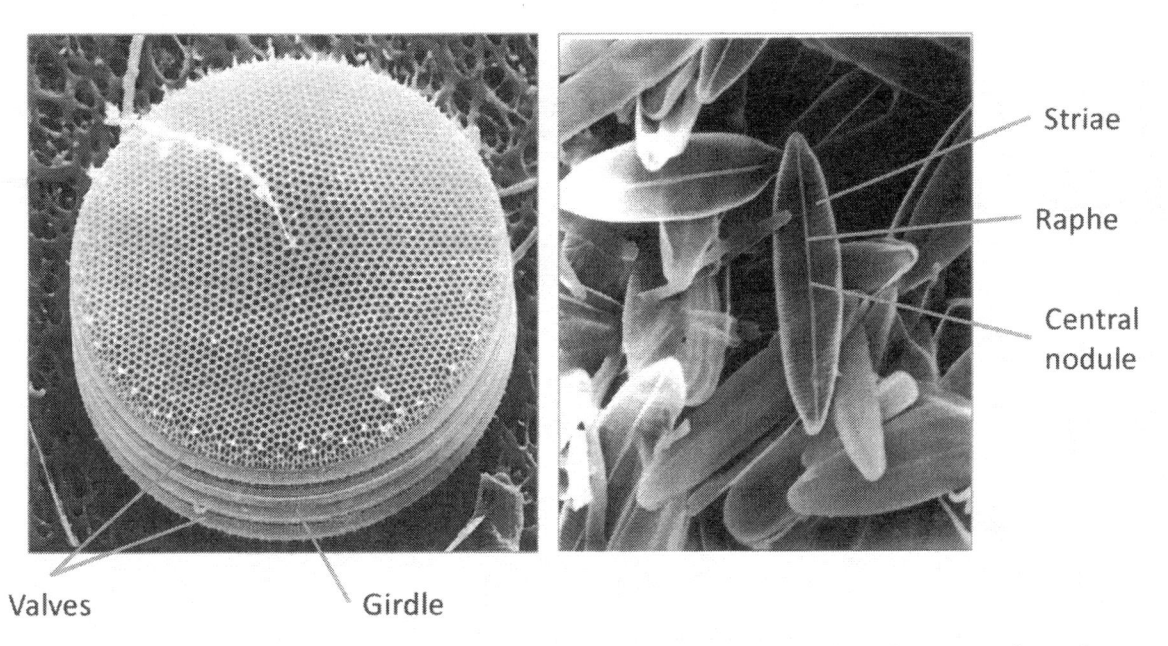

Figure 3.5. Scanning electronic micrographs of a centric frustule (left) and pennate frustule (right). (Images: Courtesy of Derek Keats of Johannesburg, South Africa, CC BY 2.0, https://creativecommons.org/licenses/ by/2.0).

Phylum Phaeophyta (word roots = gray, plant) brown algae
Brown algae may be unicellular, colonial or multicellular. Large, fleshy, multicellular forms are called **kelp**. Kelp can form underwater kelp forests or kelp beds and some species can be 45 meters long and grow as much as 60 cm day^{-1}.

1) Observe, describe and DRAW any living or preserved brown algae that are available. One species that is commonly available is *Fucus*. If *Fucus* is available, use **Figure 3.6** to help you identify what you see.

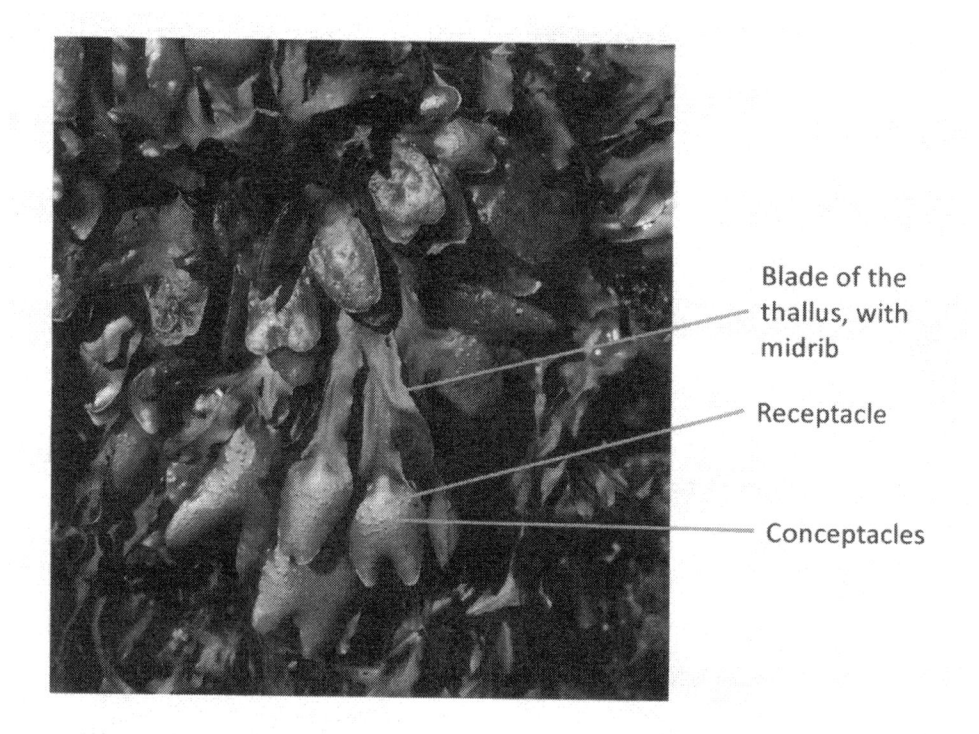

Figure 3.6. *Fucus*, a small brown kelp found in the rocky intertidal zone along the west and east coasts of North America and Europe. The body of a multicellular alga is called the **thallus**. The thallus includes the **holdfast** (structure of attachment, not shown in this photo), **stipe** and **blade**. The **receptacle** of *Fucus* is a mucus-filled structure at the tip of the blade that prevents reproductive **conceptacles** from drying out during exposure to the air during low tide. Conceptacles are the small bumps covering the receptacle, and they house reproductive structures of the organism. (Image: ARH)

2) Observe, describe and DRAW a prepared slide of a *Fucus* conceptacle. Use **Figure 3.7** to help you identify what you see.

Figure 3.7. Cross-section view of a *Fucus* conceptacle. The conceptacle produces eggs in **oogonia** and sperm on **antheridia**. The **ostiole** is the opening into a conceptacle. (Image: Modified by ARH from Curtis Clark, CC BY-SA 3.0, https://creativecommons.org/licenses/by-sa/3.0)

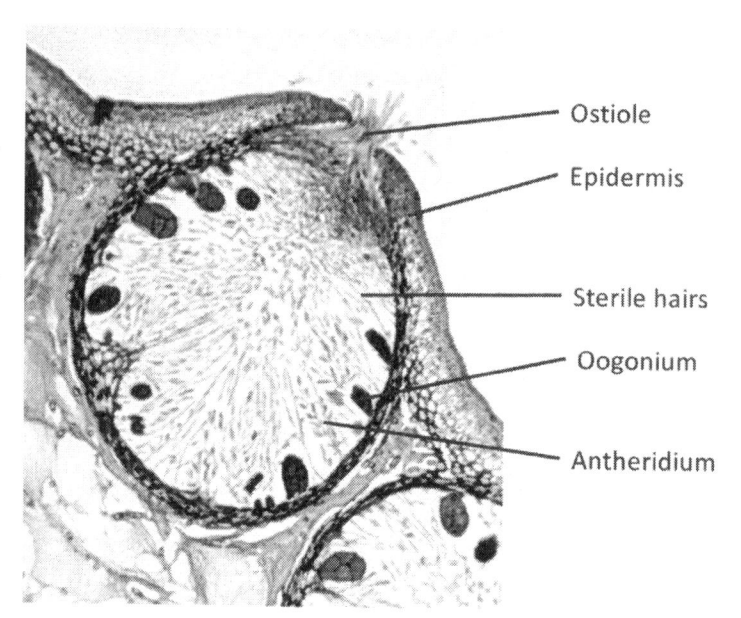

3) Observe, describe and DRAW any other available specimens of multicellular brown algae, as assigned by your instructor.

Task #3: Group Excavata (word root = out of, a hollow)

Phylum Euglenozoa (word root = good or true, socket, animal)

This is a moderate sized group of primarily freshwater protists. About two thirds of all euglenids have chloroplasts that are secondary endosymbionts of a green alga. Many euglenids are **mixotrophs**, meaning that they can function as autotrophs or heterotrophs, depending on local conditions. Euglenids are unicellular flagellates that lack a cell wall. Instead they have a series of spiral, proteinaceous strips just under the plasmalemma that form a **pellicle**. They are biflagellate, having one long emergent flagellum and one short non-emergent flagellum. Both are attached in an anterior pocket called the **reservoir** that connects to the outside by a narrower canal. There is a basal swelling on the emergent flagellum that is associated with a red-pigmented **eyespot** or **stigmata** adjacent to the reservoir. The basal swelling and stigma form a light-detecting apparatus. A **contractile vacuole** is also adjacent to the gullet and empties into the reservoir. Euglenids are most commonly found in turbid, eutrophic to hypereutrophic water or mudflats, and as a result are ecological indicator organisms of poor water quality. Euglenids can also produce harmless blooms under certain conditions.

1) Use a compound microscope to observe, describe and DRAW *Euglena*. Use **Figure 3.8** to help you identify what you see.

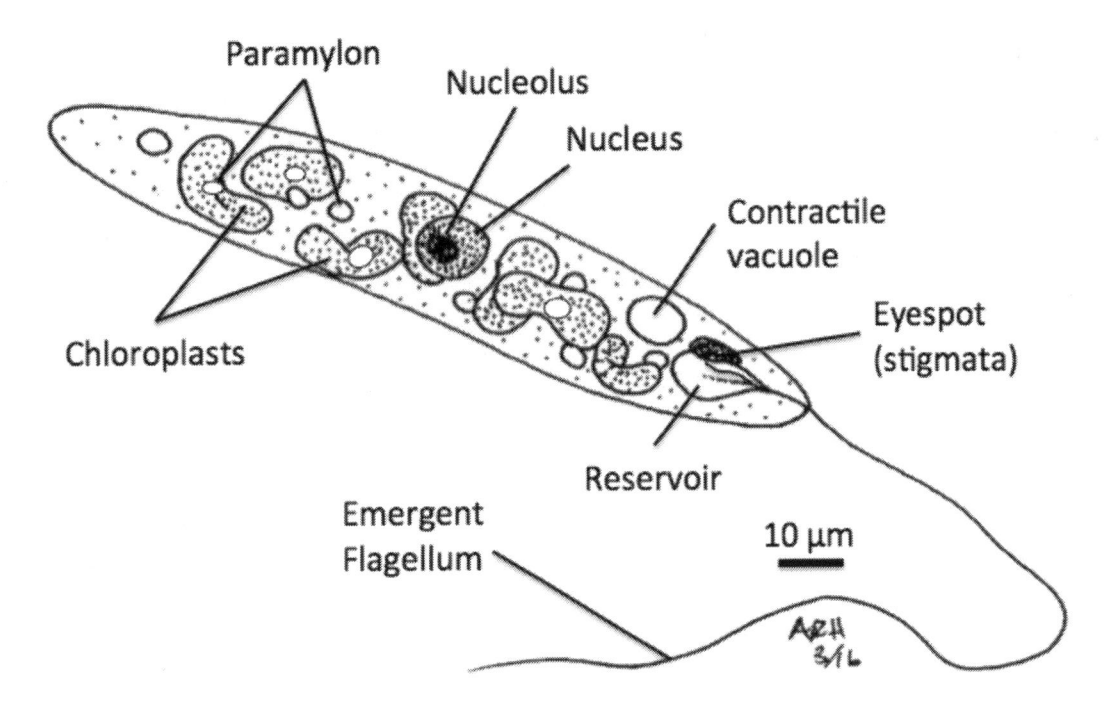

Figure 3.8. *Euglena.* **Paramylon** is a carbohydrate produced by *Euglena* to store energy. (Image: ARH)

Task #4: Group Archaeplastida (word root = ancient, folded, like)

Phylum Chlorophyta (word root = green, plant)

This is a large and diverse group of protists that occur in mainly in freshwater, but marine and terrestrial species also exist. Aquatic green algae can be found as unicellular or colony-forming phytoplankton and as multicellular seaweeds. These organisms are all autotrophic and some occur as symbionts with other organisms, notably in lichens. A few aquatic invertebrates and protists possess **zoochlorellae**, green algae that live mutualistically within their host's tissues. The chloroplasts of green algae originated by primary endosymbiosis between a heterotrophic cell and cyanobacterium. These organisms are typically bright grass-green in color. Flagellated phases occur in some species. Land plants evolved from the Chlorophyta.

1) Use a dissection or compound microscope to observe, describe and DRAW living or preserved specimens. One species you may observe is *Volvox*. *Volvox* is an extremely interesting species because it forms colonies of cells. Individual cells are flagellated and are embedded in a gelatinous matrix secreted collectively by all members of the colony. Daughter colonies are produced internally and are released when the parent colony bursts open (see **Figure 3.9**).

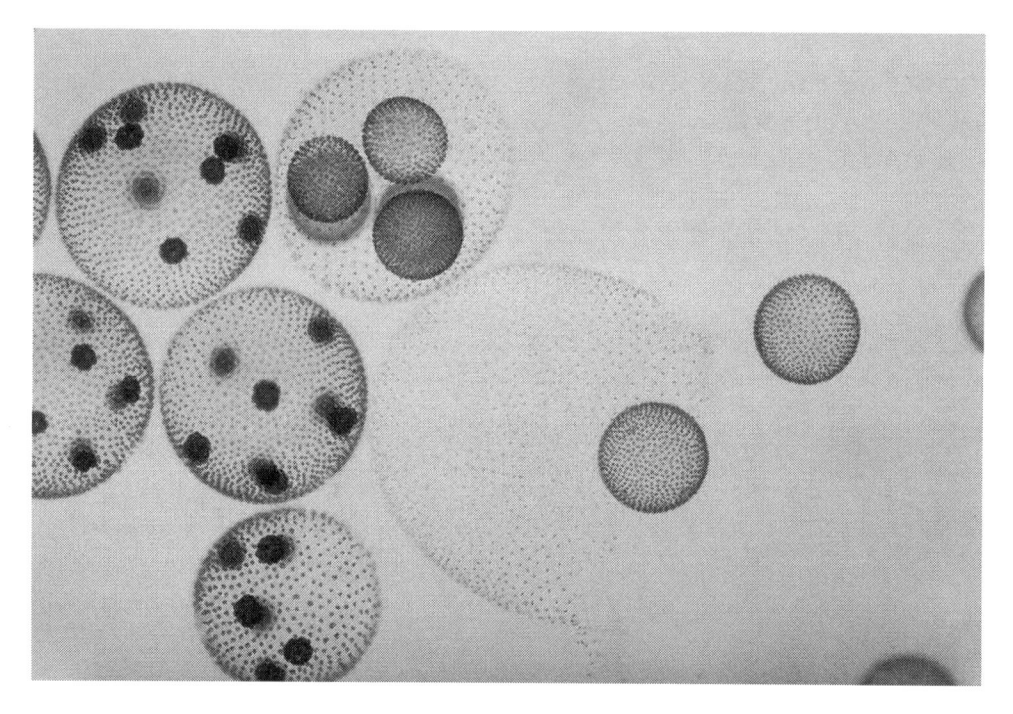

Figure 3.9. *Volvox* colonies. Colonies at the left-hand edge of the image contain daughter colonies. The colonies in the right-hand part of the image have burst open and released daughter colonies. (Image: Courtesy of Frank Fox, CC BY-SA 3.0 de, https://creativecommons.org/licenses/by-sa/3.0/de/deed.en)

Group Questions

1) Based on what you have observed in today's lab, develop an evolutionary scenario in which an animal such as a sea anemone or a coral could have formed a symbiotic relationship with zooxanthellae.
2) Based on what you have observed in today's lab, develop a scenario in which a multicellular autotroph, like kelp, could have evolved from a unicellular autotroph.
3) If your job was to evaluate the water quality of a lake, which organism from today's lab would you use as an indicator of poor water quality, and which organism from today's lab would you use as an indicator of good water quality.

Protista - Glossary

Antheridium – structure in a conceptacle that produces sperm

Blade – leaf-like organ of a fleshy alga

Cellulose test (theca) – the protective outer covering of a dinoflagellate

Central nodule – the space between two raphes in a pennate diatom

Centric diatom – diatom that is circular in shape

Cingulum – girdle around the equator of dinoflagellates, houses the transverse flagellum

Conceptacle – structure that houses reproductive cells in some fleshy algae

Contractile vacuole – organelle that carried out osmoregulation by pumping excess water out of the cell

Ectoplasm – clear zone just interior of the plasma membrane in some protists

Endoplasm – the portion of the plasma membrane that contains fluid and organelles

Epitheca – the portion of the theca of a dinoflagellate above the cingulum

Food vacuole – vacuole that bears ingested material, is also where digestion takes place

Frustule – SiO_2 outer covering of a diatom

Girdle – wall that runs around the equator of diatom frustules

Gullet – base of the oral groove, where food is placed in vacuoles

Holdfast – attachment organ of a fleshy alga

Hypotheca – the portion of the theca of a dinoflagellate below the cingulum

Kelp – large, fleshy brown algae

Lobopodia (pseudopodia) – broad pseudopods, like those of amoeba

Longitudinal flagellum – flagellum that is housed in the sulcus of a dinoflagellate

Macronucleus – nucleus that controls non-reproductive functions of the cell

Micronucleus – nucleus that is involved in sexual reproduction

Mixotroph – organism that can function as a heterotrophic or an autotrophic as needed

Oogonium – structure in a conceptacle that produces or houses eggs

Oral groove – indentation into the plasma membrane of ciliates where food is captured and placed in a food vacuole

Ostiole – opening into a conceptacle

Paramylon – a carbohydrate storage molecule, similar to starch

Pellicle – protective proteinaceous strips in the plasma membrane of *Euglena*

Pennate diatom – diatom that has an elongate shape

Phytoplankton – floating photosynthetic organisms

Pseudopodia – extensions of the plasma membrane

Raphe – slit in the frustule of a diatom, possibly the site of locomotion

Receptacle – structure of an alga that bears conceptacles

Red tide – population explosion of dinoflagellates that release toxins into the water, sometimes killing fishes and making clams dangerous to eat.

Reservoir (of *Euglena*) – fluid-filled compartment that houses the flagellae of *Euglena*

Stigma – pigmented patch that casts shade on the reservoir, which is light-sensitive

Stipe – stem-like organ that supports the blade of a fleshy alga

Striae – ornamental pattern radiating from the center of a diatom

Sulcus – longitudinal groove that extends from the cingulum to the base of a dinoflagellate, houses the longitudinal flagellum

Thallus – body of a fleshy alga

Theca – the outer covering of dinoflagellates

Thecal plate – one piece of the theca of a dinoflagellate

Transverse flagellum – flagellum that encircles the equator of a dinoflagellate

Trichocyst – defensive organelle that fires like a harpoon when stimulated

Valve – one piece of outer covering, like a shell

Zoochlorellae – unicellular green algae that are mutualistic endosymbionts, mainly in freshwater

Zooxanthellae – any of several species of dinoflagellates that are mutualistic endosymbionts with various animals, mainly in marine habitats

Chapter 4: Plantae - Nonvascular Plants

There is compelling evidence that plants evolved from green algae (Chlorophyta) and that the Charophyceae, a group of commonly called the stoneworts, is the sister taxon to Kingdom Plantae. **Table 4.1** provides an overview of similarities and differences between these groups.

Table 4.1. Similarities and differences between Charophyceae and Plantae.

Characteristic	Charophyceae and Plantae shared character	Charophyceae	Plantae
Chlorophyll a & b, carotenoids, xanthophylls	Yes	-----	-----
Starch stored in plastids as the food reserve	Yes	-----	-----
Peroxisome structure and function	Yes	-----	-----
Cell wall composition and cellulose synthesis complexes	Yes	-----	-----
Phragmoplast formation during cytokinesis	Yes	-----	-----
Flagellum ultrastructure	Yes	-----	-----
Apical meristem growth	-----	No	Yes
Multicellular gametangia	-----	No	Yes
Retention of embryo in archegonium	-----	No	Yes
Desiccation resistant cuticle	-----	No	Yes
Stomata/pores for gas exchange	-----	No	Yes
Alternation of generations life cycle	-----	A few (some green algae)	Yes
Alternation of heteromorphic generations	-----	No	Yes

The **alternation of generations life cycle** characteristic listed in **Table 3.1** is of particular interest, evolutionarily. In the plant life cycle, also known as the **diplohaplontic life cycle**, two separate phases exist. In one phase a multicellular diploid **sporophyte** produces haploid spores by meiosis. In the next phase, each haploid **spore** can germinate independently and develop into a multicellular haploid **gametophyte**. The haploid gametophyte produces haploid **gametes** by mitosis. Complementary gametes meet and undergo syngamy to produce a diploid **zygote** that will develop into a multicellular diploid sporophyte and you are back where you started (see **Figure 4.1**). In addition, members of Kingdom Plantae exhibit an alternation of **heteromorphic** generations. This means that the multicellular sporophyte and gametophyte stages have different forms or shapes. Most green algae, on the other hand, have a **haplonitic life cycle**. This means that their only multicellular stage (if they have one) is the haploid gametophyte. The gametophyte produces gametes by mitosis and they undergo **syngamy** to produce a **zygote**, the only diploid life stage in this life cycle. The zygote then undergoes **meiosis** right away to produce haploid **spores** that germinate independently and grow into new multicellular, haploid **gametophytes**. Some green algae do produce alternating multicellular **sporophyte** and multicellular **gametophyte** stages, but when they do, these stages are **isomorphic**, that is, the sporophytes and gametophytes look the same.

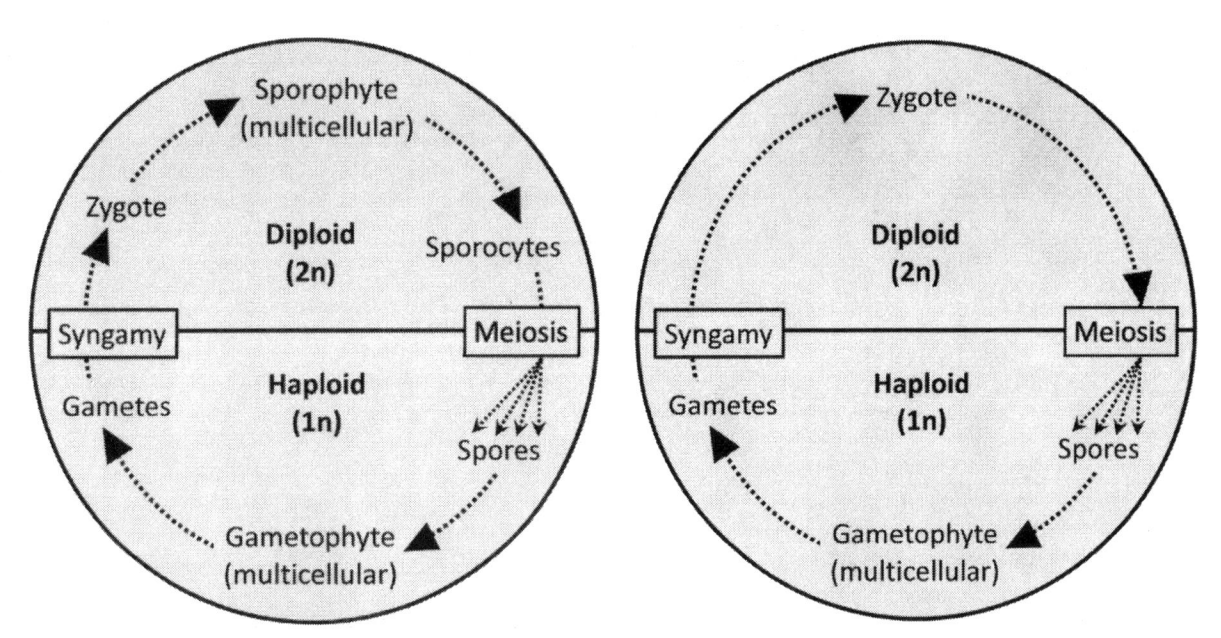

Figure 4.1. Left: Diplohaplontic (alternation of generations) life cycle of plants. **Right**: Haplontic life cycle of most green algae. Terms in boxes indicate processes by which organisms move from one life phase to the next. (Images: ARH)

Historically, plants were divided into two major taxonomic groups. One group, the vascular plants, the **Tracheophyta**, are those that have the plant vascular tissues xylem and phloem. The other group, the non-vascular plants, the **Thallophyta**, lack vascular tissues and the body is referred to as a **thallus**. The term thallophytes is no longer used, because all of the non-vascular plants were assigned to Phylum Bryophyta and were called bryophytes. Phylogenetic analysis of the old Phylum Bryophyta, however, revealed that it is not monophyletic, and contains three distinct lineages: the liverworts (Phylum Marchantiophyta), the hornworts (Phylum Anthocerotophyta) and the mosses (Phylum Bryophyta), as shown on the plant phylogenetic tree in **Figure 4.2**.

There is an interesting evolutionary trend in the plant kingdom concerning its alternation of generations lifecycle. In non-vascular plants, the gametophyte is the dominant phase, the sporophyte is reduced, and its spore-producing structure is **monosporangiate** (unbranched) and wholly dependent on the gametophyte. In other words, as someone once famously quipped, "In the bryophyte, the sporophyte is a parasite of the gametophyte." Conversely, in tracheophytes the sporophyte phase is dominant and is either independent of a much-reduced gametophyte or produces a tiny gametophyte internally. For example, in seed plants the gametophyte develops within the tissues of the sporophyte and is wholly dependent on sporophyte tissue for survival. In addition, sporophytes in this group are **polysporangiate** (branched).

```
                    BRYOPHYTES (or "THALLOPHYTES")
                         (non-vascular plants)
CHAROPHYTES    MARCHANTIOPHYTA   ANTHOCEROTOPHYTA   BRYOPHYTA    TRACHEOPHYTES
(green algae)    (liverworts)        (hornworts)     (mosses)    (vascular plants)
```

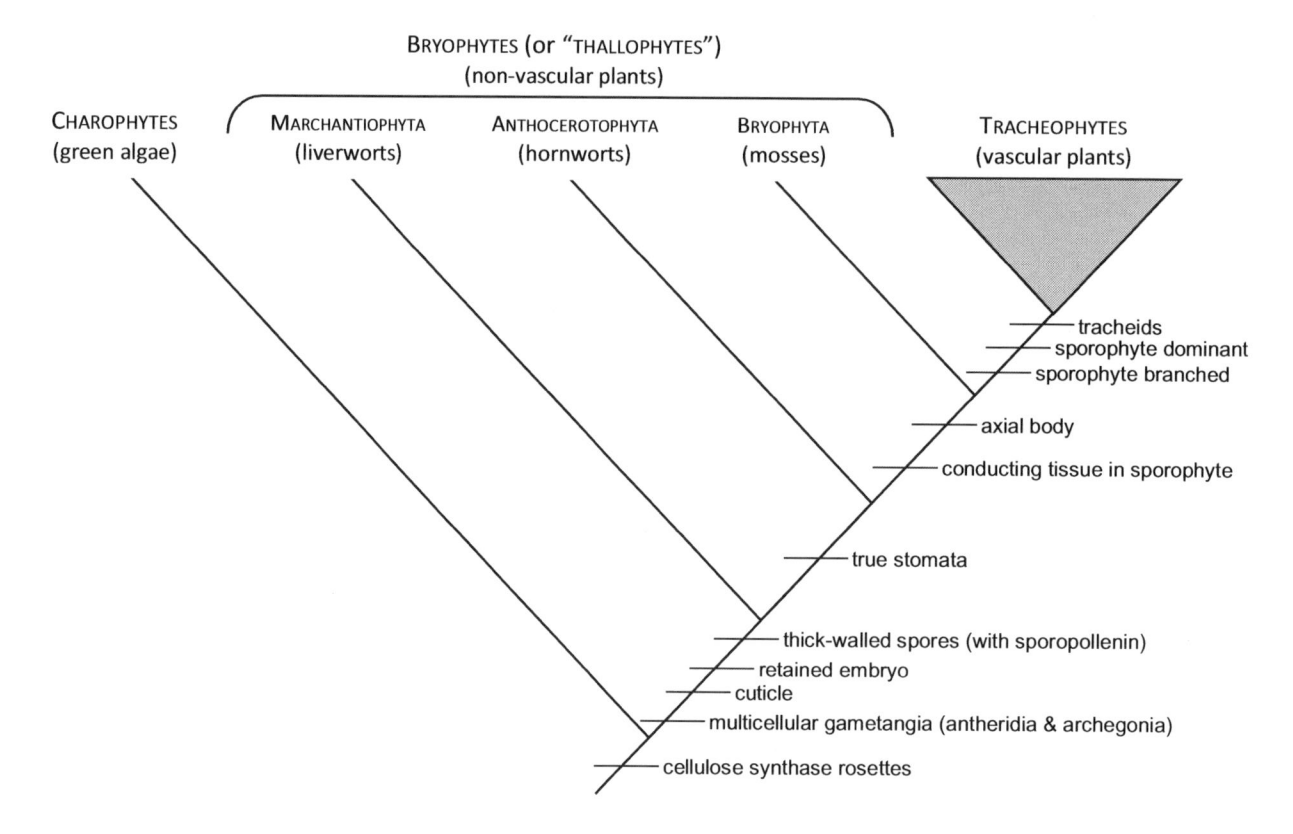

```
                                                          ─ tracheids
                                                         ── sporophyte dominant
                                                        ─── sporophyte branched
                                                  ── axial body
                                             ── conducting tissue in sporophyte
                                       ── true stomata
                                  ── thick-walled spores (with sporopollenin)
                                 ── retained embryo
                                ── cuticle
                               ── multicellular gametangia (antheridia & archegonia)
                              ── cellulose synthase rosettes
```

Figure 4.2. One proposed phylogeny of non-vascular plants. Note that the formerly accepted Thallophyta is not monophyletic. (Image courtesy of Dr. Gary Baird, Dept of Biology, BYU-Idaho)

Task #1: Phylum Marchantiophyta (word roots = Marchant's plants; Marchant was a French botanist)

In **Phylum Marchantiophyta** (**liverworts**) the thallus is often flattened and **dichotomously branched** to form distinct lobes. So, why are they commonly called liverworts? The shape of the thallus reminded early herbalists of the lobes of a liver, and hence the name liverwort (the term *wort* is derived from the Old English *wyrt*, meaning herb). Liverworts can be separated into two distinct groups: 1) thalloid liverworts – those with a typical thallus, and 2) leafy liverworts – those in which the thallus is divided into leaf-like segments. There are approximately 7200 species of liverworts worldwide, the majority of these being tropical, leafy liverworts.

In liverworts, the upper surface of the gametophyte has **pores**. Unlike stomata, these pores cannot close because there are no guard cells. As a result, these plants are constantly transpiring (releasing water). Also, on the upper surface of the thallus there may be **gemma cups**. These structures contain **gemmae**, which are small multicellular, disc-shaped pieces of thallus that are produced asexually. When gemmae are dislodged each one can grow into a new thallus.

As in all bryophytes, liverworts lack true roots. Instead, the undersurface of the thallus produces numerous hair-like **rhizoids** that are used to anchor the plant to the substrate.

Female and male liverwort gametes are often produced on separate thalli, with the **archegonia** and **antheridia** produced on upright stalks called **archegoniophores** and **antheridiophores**, respectively. These are small, umbrella-like structures. Archegonia are produced on the underside of the archegoniophore but **antheridia** are produced on the upper surface of the antheridiophore. Raindrops splash flagellated sperm from an antheridiophore onto an archegoniophore. The sperm must then swim to an archegonium where syngamy occurs. The resulting **zygote** develops into an embryo which is housed within the remains of the archegonium, now called a **calyptra**. This embryo then becomes a microscopic sporophyte that consists of a **foot** that connects the sporophyte to the gametophyte, a short stalk-like **seta**, and sporangium called the **capsule**. Spores are produced in the sporangium and are released to grow into new gametophyte thalli.

1) Use a dissection microscope to observe a living liverwort gametophyte of *Marchantia*. Study a model of a liverwort if live material is not available.
    a. Observe, describe and DRAW the **thallus** of a liverwort. Label the **thallus**, **rhizoids**, **pores** and a **gemma cup**.
    b. DRAW a **gemma cup** in detail with its **gemmae**. Refer to the center image of **Figure 4.3** to help you identify these structures.
2) Use a compound microscope to describe a prepared slide of *Marchantia* antheridium. DRAW and label the **antheridiophore** (you will only see a small portion of it) and an **antheridium**. Use **Figure 4.3** upper right-hand image to help you identify what you see.
3) Use a compound microscope to observe and describe a prepared slide of a *Marchantia* archegonium. DRAW and label the **archegoniophore** (you will only see a small portion of it) and an **archegonium**. Label the **venter** and **neck of the archegonium**. Use the upper left-hand image of **Figure 4.3** to help you identify what you see.
4) Use a compound microscope to observe and describe a prepared slide of a *Marchantia* sporophyte. DRAW and label the **sporophyte**, including the **foot**, **seta**, and **capsule**. Also find and label the **spores**, **elaters**, and **calyptra**. Refer to **Figure 4.4** to help you identify what you see.

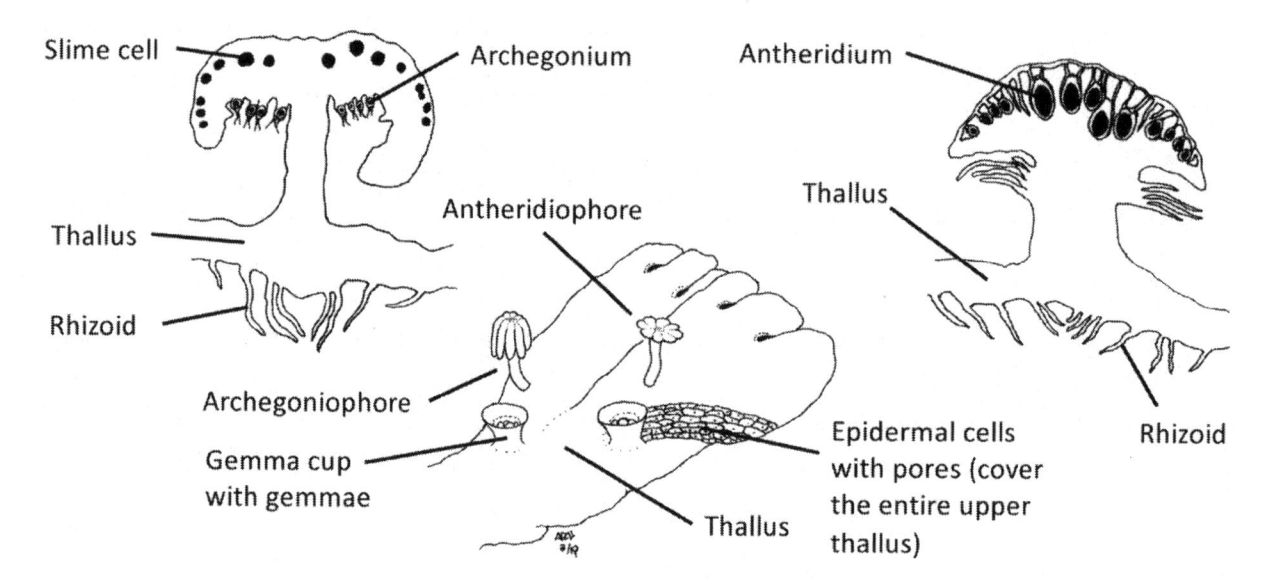

Figure 4.3. The liverwort *Marchantia*: thallus and structures (center image), cross-section of archegoniophore and archegonia (upper left-hand image) and cross-section of antheridiophore and antheridia (upper right-hand image). Note: Only a few epidermal cells are shown on the image of the thallus, and no cells of the archegoniophore and antheridiophore are shown in order to highlight important structures. (Images: ARH)

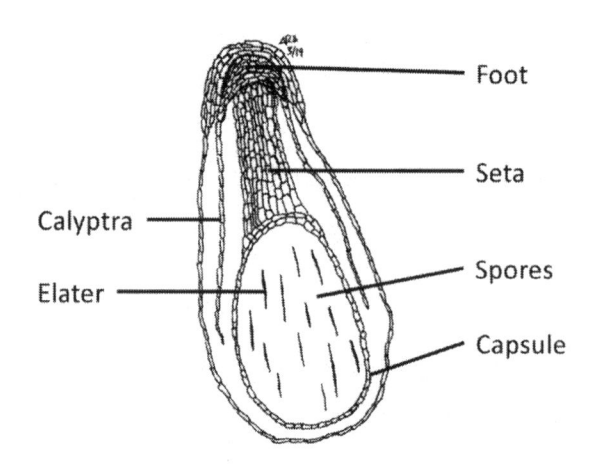

Figure 4.4. Sporophyte of *Marchantia*. Note: Spores mixed with elaters fill the entire capsule but spores are not drawn so elaters would not be obscured. (Image: ARH)

Task #2: Phylum Anthocerotophyta (word roots = flower, horned, plant)

Phylum Anthocerotophyta (hornworts) is a small group with about 200 species. Hornworts typically produce a small, roundish gametophyte thallus that is obscurely dichotomously branched and has true stomata on the upper surface. Hornworts get their name from the growth of sporophyte stage that has a horn-like appearance as it grows upward from the gametophyte base. **Archegonia** and **antheridia** are produced directly on the surface of either the same gametophyte thallus or on separate gametophyte thalli.

The sporophyte consists of a foot and an elongated cylindrical capsule, which grows from a basal meristem and splits longitudinally when mature to release the spores.

1) Specimens of this group are notoriously difficult to obtain. If specimens are available, follow your instructor's directions to observe, describe and DRAW them.

Task #3: Phylum Bryophyta (word roots = moss, plant)

Phylum Bryophyta, mosses, is the largest group of bryophytes with over 12,000 described species. Unlike liverworts and hornworts, mosses are not thalloid. Instead the gametophyte has a stem-like **central axis** that bears small leaf-like **blades**. **Rhizoids** are produced at the base of the stem and anchor the gametophyte. As in liverworts, the gametophyte may produce **archegonia** or **antheridia** or both at the apex of the stem. The **sporophyte** consists of a **foot**, elongated **seta** (usually), and terminal **capsule** in which spores are produced. The **sporophyte** grows out of the top of the female **gametophyte**. The sporophyte's capsule is typically covered by a structure called the **calyptra** which eventually falls off, and has a lid-like **operculum**. When the capsule is mature, the operculum opens, sometimes abruptly, often revealing a ring of teeth called the peristome which aids in spore dispersal. Mosses possess rudimentary conducting cells in both the gametophyte and sporophyte; these however, are not the same as vascular tissues of tracheophytes.

1) Observe and describe living material or a model of a bryophyte gametophyte with attached sporophyte. If you are studying fresh material, use a dissection microscope or hand lens to observe a gametophyte with attached sporophyte. DRAW and label the **gametophyte, sporophyte,** and **seta, capsule, operculum** and **calyptra** of the **sporophyte**. On the **gametophyte**, label the **blades, central axis** and **rhizoids**. Use **Figure 4.5** to help you identify what you see.

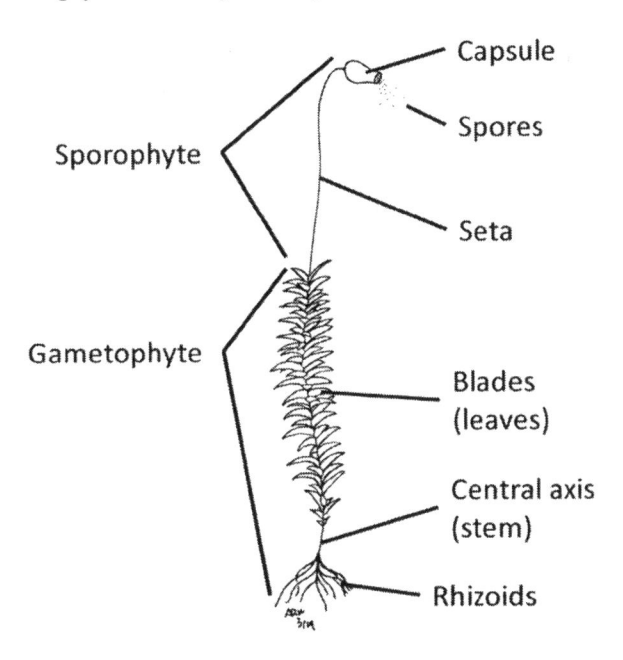

Figure 4.5. A bryophyte gametophyte with sporophyte. (Image: ARH)

2) Use a dissection microscope as you observe, describe and DRAW a **protonema**, the juvenile stage of a bryophyte. The protonema will look like a filamentous green alga. As it matures, it will develop into a mature **gametophyte**.
3) If available, study a blade from the haircap moss *Polytrichum*. Haircap mosses have a distinctive leaf anatomy. Examine the upper surface of a leaf and notice it is striated – has many ridges or grooves. This is due to closely spaced lamellae (thin plates) running the length of the leaf. Gently scrape the surface of the leaf to dislodge the lamellae so you can detect them more easily. DRAW a leaf and label the lamellae.
4) Use a compound or dissection microscope as you observe, describe and DRAW prepared slides of a longitudinal section of *Polytrichum* capsules. Dried capsules may also be available for examination. The operculum is the cap-like structure at one end and the seta is at the other end. DRAW and label the **capsule, operculum, seta** and **spores**. Refer to **Figure 4.6** to help you identify what you see.

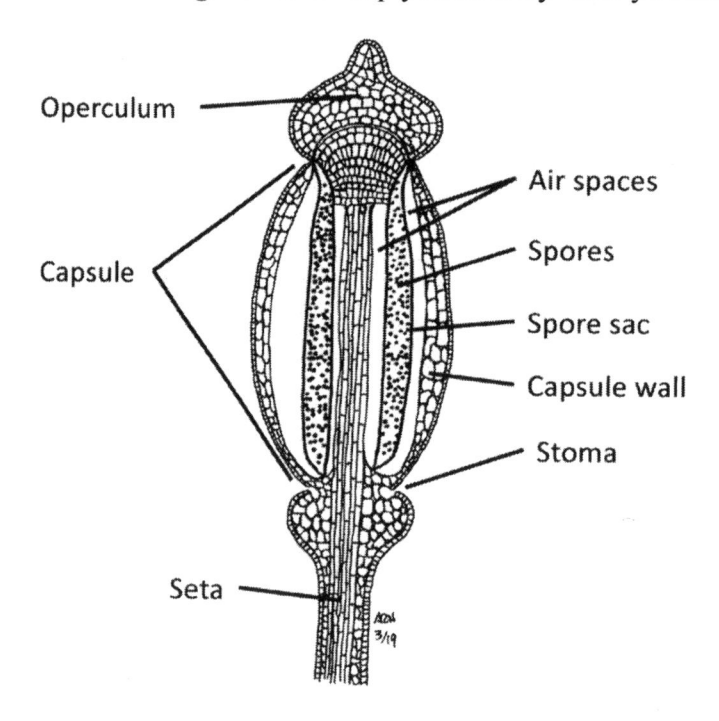

Figure 4.6. Longitudinal section of a sporophyte capsule of the bryophyte *Polytrichum*. (Image: ARH)

5) Observe, describe and DRAW living material of the bryophyte, *Sphagnum*. Sphagnum mosses have a distinctive leaf anatomy. Make a wet-mount slide of a sphagnum moss leaf and examine it under a compound microscope. Notice the pattern of living (green) cells and dead (clear) cells in the leaf. DRAW a leaf and label the living cells and non-living cells.

Group Questions

1) Develop a scenario that explains how a multicellular green alga could have evolved into a non-vascular plant with a diplohaplontic life cycle.
2) Describe the impact of environmentally resistant spores of non-vascular plants to the evolution of land plants.
3) Explain why the now defunct taxon Thallophyta (**Figure 4.2**) is no longer accepted as a viable taxon. If needed, refer to Chapter 1 (Phylogenetics) to answer this question.

Non-vascular plants – Glossary

Alternation of generations life cycle – see diplohaplontic life cycle

Antheridium/antheridia (plural) – structure that produces sperm

Antheridiophore – structure (stalk) that supports/houses antheridia

Archegonium/archegonia (plural) – structure that produces eggs

Archegoniophore – structure (stalk) that supports/houses archegonia

Blade – non-vascular leaf-like structure

Calyptra – remains of an archegonium that houses a developing zygote

Capsule (sporangium) – structure in which spores are produced

Central axis – non-vascular stem-like structure

Dichotomous branching – when a structure branches in two

Diplohaplontic life cycle – life cycle containing a multicellular diploid sporophyte phase and a multicellular haploid gametophyte phase

Elater – structures in the capsule of a sporophyte that aid in releasing spores by coiling and uncoiling

Foot of a sporophyte – structure that connects a developing sporophyte to a gametophyte

Gamete – reproductive haploid cell that is used to carry out syngamy

Gametophyte – multicellular life stage that produces gametes

Gemma cups – cup shaped structures that produce multicellular gemmae, multicellular structures produced by cloning that can dislodge and produce a new thallus

Gemmae – multicellular structures produced by cloning in gemma cups, used for asexual dispersal of the thallus

Haplontic life cycle – life cycle in which the only multicellular phase is haploid

Heteromorphic – structures look different than each other

Isomorphic – structures look the same, e.g., sporophyte and gametophyte stages in some algae

Meiosis – specific type of cell division in which a diploid cell produces four haploid cells

Monosporangiate – spore producing structure that is not branched (in non-vascular plants)

Neck of the archegonium – canal through which sperm must pass to reach the egg

Operculum – structure that covers the top of a sporangiophore and it opens, sometimes explosively, during spore release

Polysporangiate – spore producing structure that is branched (in vascular plants)

Pores (central pores) – openings through cell walls of epidermal cells of some non-vascular plants that facilitate gas exchange

Protonema – thread-like mass of haploid cells produced by a germinating spore gives rise to the gametophyte

Rhizoid – root-like structure used to anchor non-vascular plants to the substrate

Seta – stalk that supports or will eventually support the capsule of a sporophyte

Slime cells – slime cells absorb and retain water

Sporangium – structure in which spores are produced

Spore – single cell that can geminate and develop into a multicellular body without undergoing syngamy

Sporocyte – cell that produces spores by meiosis

Sporophyte – multicellular life stage that produces spores

Stoma – opening that allows gas exchange

Syngamy – fusion of two complementary reproductive cells (usually egg and sperm), sometimes referred to as fertilization

Thallophyta – non-vascular plants, plants that do not have xylem and phloem

Thallus – multicellular phase of a non-vascular plant

Tracheophyta – all plants that have vascular tissues, e.g., xylem and phloem

Venter – swollen base of an archegonium that houses the egg

Zygote – product of syngamy

Chapter 5: Plantae - Non-flowering Vascular Plants

The appearance of vascular tissue was a major step in plant evolution. Vascular tissue consists of **phloem**, the tissue that carries photosynthate (mainly sugars), and **xylem**, the tissue that carries water and dissolved minerals. A distinctive feature of xylem is the presence of **tracheids**. Tracheids are specialized cells that possess a thickened secondary cell wall that has been stiffened by the addition of **lignin**. The secondary cell wall of tracheids develops at least initially as rings or spirals and are reminiscent of the human trachea, hence their name. Vascular plants containing tracheids in their conducting tissues are the **Tracheophyta**, while those that lack such cells are the **Thallophyta**, non-vascular plants, covered in Chapter 4.

All plants exhibit an alternation of generations. In the **Thallophyta** the haploid **gametophyte** phase is the dominant generation, while in the **Tracheophyta** the diploid **sporophyte** phase is dominant.

The evolutionary trend in **tracheophytes** is a pattern of reduction in the size of the **gametophyte** and increase in the size of the **sporophyte**. In some tracheophytes the **spores** are **homosporous** (i.e., all the same size), while in others they are **heterosporous** (i.e., spores of different sizes). Heterosporous plants have structures called **microsporangia** that produce small spores called **microspores**, and other structures called **megasporangia** that produce large spores called **megaspores**. Generalized life cycles of homosporous and heterosporous plants are presented in **Figure 5.1**.

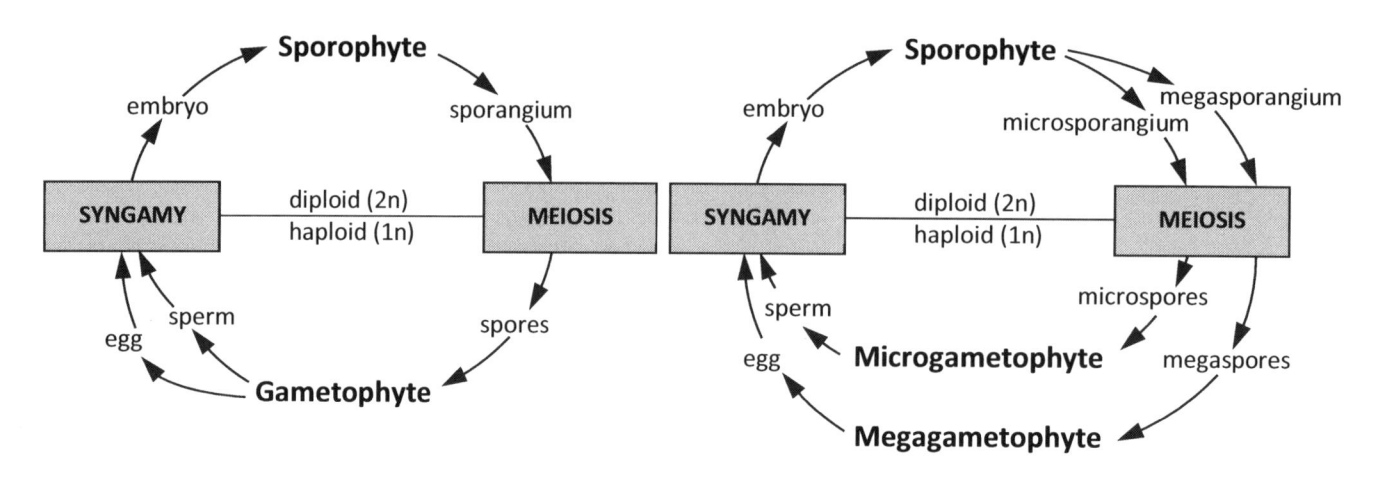

Figure 5.1. Left: Homosporous life cycle. **Right:** Heterosporous life cycle. (Images courtesy of Dr. Gary Baird, Dept. of Biology, BYU-Idaho)

<u>Task #1:</u> **Phylum Monilophyta (word roots = necklace, plant)**

Phylum Monilophyta includes ferns and their close relations. The name Monilophyta refers to the appearance of the **stele**, the ring of connective tissue found in the stems of many species in this group. The stele is reminiscent of a necklace when the stem is viewed in cross-section. Monilophytes typically produce **euphylls** (leaves) that develop from **apical** or **marginal meristems**.

1) Whisk ferns (Psilopsida) are a small group of tropical plants that have highly reduced features (**Fig. 5.2**). For example, in *Psilotum* there are no roots, although rhizoids are produced, and there are **enations** instead of leaves. The plant body consists of a branching **rhizome**, and the branches form photosynthetic and fertile aerial stems. Whisk ferns were once thought to be descended directly from the first vascular plants, but we now know that this is not correct.
    a. Observe, describe and DRAW one of these if specimens are available.

Figure 5.2. Psilotum whisk fern. The small thorn-like bumps on the stems are enations. (Image: ARH modified photograph by Jon Houseman under Creative Commons Attribution-Share Alike 4.0 International license)

2) Ophioglossid ferns (Ophioglossales). This small group of ferns share a number of features in common with the whisk ferns, such as a reduction in root structure and features of the sporangia (**Fig. 5.3**).
    a. Observe, describe and DRAW one of these if specimens are available.

Figure 5.3. Ophioglossid fern, Adder's tongue fern. (Image: ARH modified photograph by Netha Hussain under Creative Commons Attribution 2.0 Generic license)

3) Horsetails (Equisetopsida) or scouring rushes, are a small group of plants with a worldwide distribution. These plants differ from all other members of this phylum in having jointed stems with hollow internodes, and reduced scale-like leaves that are whorled at the nodes (**Fig. 5.4**). Like whisk ferns, horsetails have perennial rhizomes that produce annual aerial stems.
    a. Observe, describe and DRAW one of these if specimens are available.

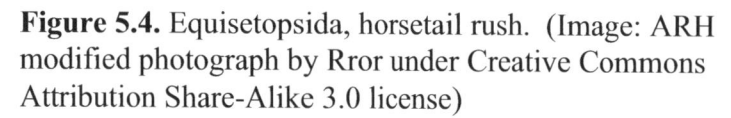

Figure 5.4. Equisetopsida, horsetail rush. (Image: ARH modified photograph by Rror under Creative Commons Attribution Share-Alike 3.0 license)

4) Leptosporangiate ferns (Polypodiopsida) (**Fig. 5.5**). This is the largest group of ferns of the monilophyte taxa, with over 10,000 described species. Only the flowering plants (angiosperms) have more species. More than 75% of fern species are tropical. A defining feature of these ferns is the presence of **leptosporangia**. These spore-producing structures differ from eusporangia found in all other vascular plants, because **leptosporangia** originate from a single cell, develop a sporangial wall only one cell layer thick, have an **annulus** on the sporangium's **capsule** that aids in dispersing the spores, and produce a reduced number of spores – fewer than 100 per sporangium. Sporangia are typically clustered into **sori** that are located on the undersurface of **fronds**, i.e., **euphylls** or leaves. Sori may be regularly spaced or sporadically arranged. Sori can also be covered by an **indusium**, a flap of tissue that protects the developing sporangia. Ferns produce small, heart-shaped gametophytes called **prothallia**. The **prothallium** is photosynthetic, grows on the soil surface, and produces both **archegonia** and **antheridia**. Fern sperm are multi-flagellated and swim through a film of water to reach an egg in an archegonium.

Figure 5.5. Leptosporangiate fern. (Image: ARH modified photograph by Jesmond Dene under GNU Free Documentation license v. 1.2)

    a. Become familiar with the fern life cycle (**Fig. 5.6**).
        i. The mature diploid **sporophyte** produces many clusters of diploid **leptosporangia** called **sori** on the underside of its **fronds**.
        ii. Diploid cells in each leptosporangium undergo meiosis to produce haploid **homosporous spores**.
        iii. **Spores** are released and disperse into the environment. Each haploid **spore** is capable of germinating and growing into a heart-shaped, haploid **prothallium (gametophyte)**.
        iv. Each prothallium can produce many **archegonia** and **antheridia**. An archegonium produces only one egg, while each antheridium yields many flagellated sperm.
        v. When eggs are mature, **archegonia** open, allowing **syngamy** when flagellated sperm are able to contact the egg, creating a diploid **zygote**.

vi. The **zygote** develops in place, and produces a diploid **sporophyte** that grows directly out of the **prothallium**.

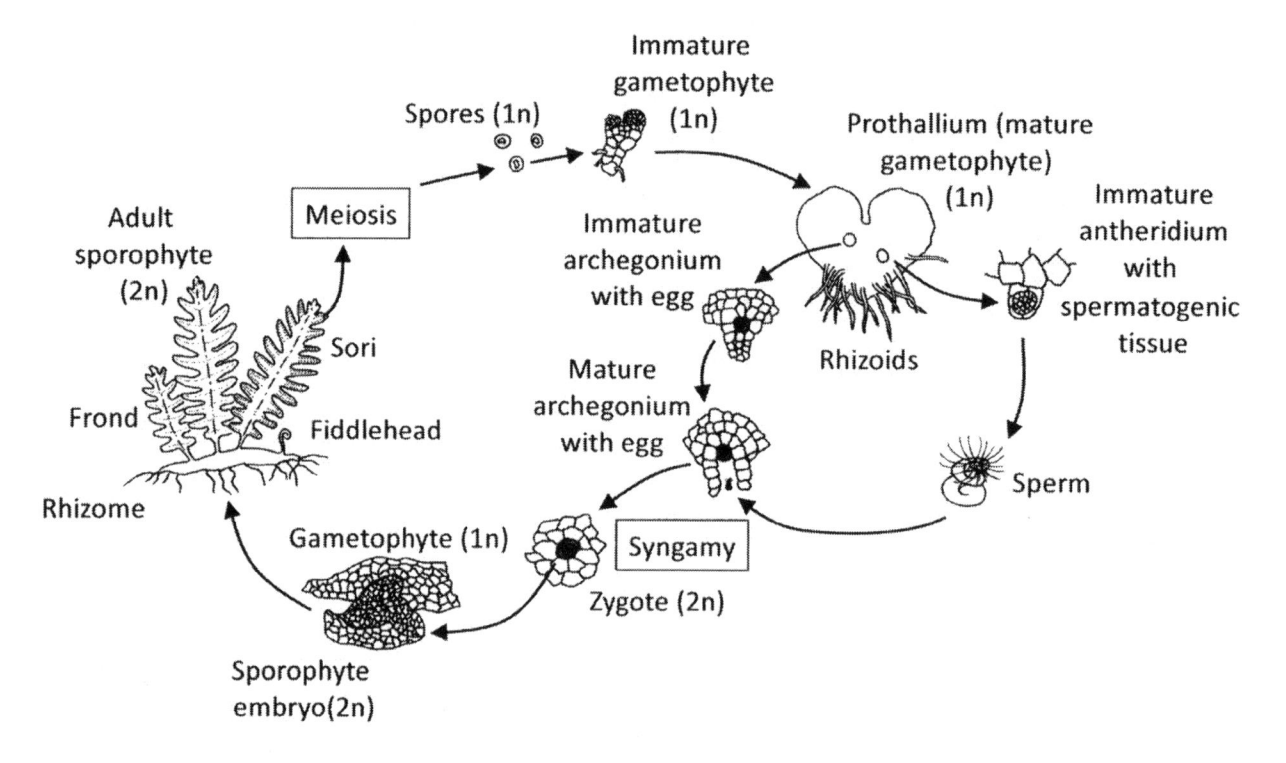

Figure 5.6. The leptosporangiate fern life cycle. (Image: ARH)

b. Observe, describe and DRAW any available living material of ferns. Observe the sporophytes and their **rhizomes**. Look on the underside of **fronds** to see if **sori** are present.
c. DRAW an example of a young frond, called a **fiddlehead**, showing **circinnate vernation**.
d. Use a dissection microscope or hand lens to study a sorus. Note the presence or absence of an **indusium**. Keep looking until you find a frond with indusia. An indusium is a thin layer of tissue that covers a sorus. If you can see a **sporangium** then an **indusium** is not present. DRAW the pattern of sori on the frond.
e. DRAW a **leptosporangium**. Label the **sorus, leptosporangia, annulus** and **indusium**. Refer to **Figure 5.7** to help you identify what you see.

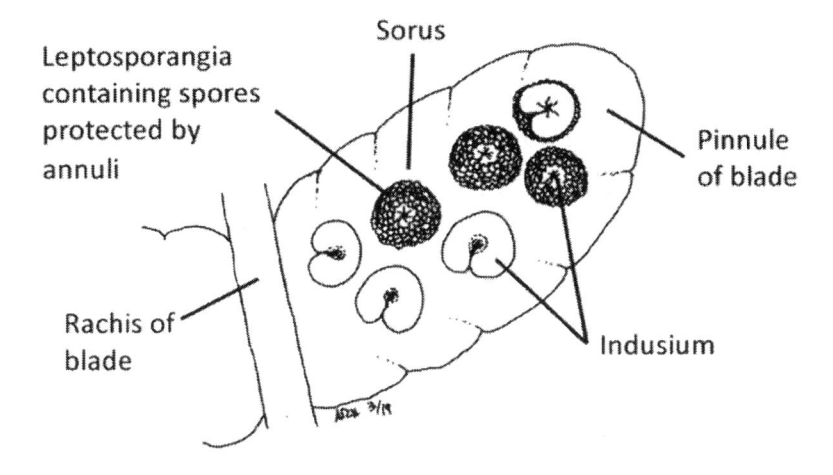

Figure 5.7. Underside of a fern **sporophyll** (blade/frond) bearing **sori**. Each **leptosporangium** contains spores and an outer layer of cells called the **annulus**. (Image: ARH)

    f. Use a compound microscope to examine a prepared cross-section slide of a fern frond (blade). Locate a **sorus** where the umbrella-like **indusium** is visible. Observe, describe and DRAW a sorus and label the **sporophyll** (frond), **sorus, leptosporangium, annulus** and **indusium**. Use **Figure 5.8** to help you identify what you see. Be advised that the slide you observe will not show the sporangia and their stalks as clearly as this idealized drawing.

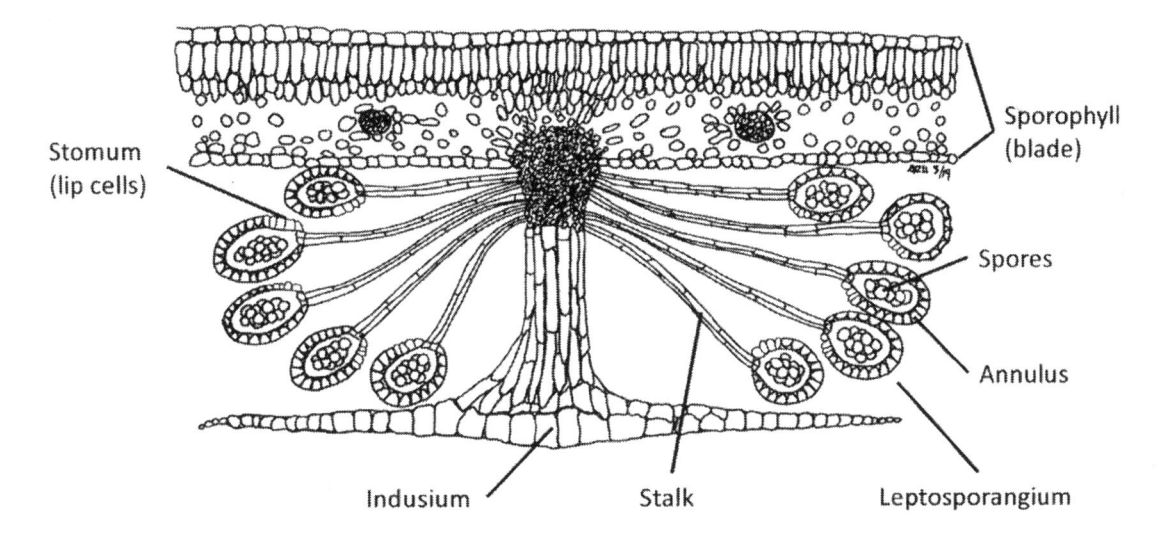

Figure 5.8. Cross-section through a sorus with an indusium. (Image: ARH)

    g. Observe, describe and DRAW a fern **prothallium**. Prepare a wet mount of living fern prothallia, and depending on its size, observe one using a hand lens, dissection microscope or compound microscope. Look for the classic heart-shaped form, **rhizoids** that originate near the pointed end of the prothallium, and brownish **archegonia** near the notch. Watch for swarming sperm cells which may be mistaken for fast moving protists. Label the following on your drawing: **prothallium**, **rhizoids**, **archegonia**, and **antheridia** though antheridia may not be readily visible.

h. Make a wet-mount slide of a fern prothallium with an attached young sporophyte and then use a dissection or compound microscope to observe, describe and DRAW it. The young sporophyte will be growing out of the archegonium where the egg was fertilized. Be sure to label the **prothallium** and **sporophyte embryo**. Refer to the life cycle in **Fig. 5.6** for a representation of what this might look like.

Task #2: Gymnosperms

Spermatophytes are plants that produce seeds. The development of seeds was another important step in plant evolution. Seeds are environmentally resistant propagules that can survive adverse conditions. This is most likely one significant reason why seed plants dominate over non-seed plants today.

There are five extant groups of spermatophytes. Four of these are collectively referred to as **Gymnosperms** (*gymnos*, naked, and *sperma*, seed). Their seeds are naked or exposed to the environment. The other group (covered in Chapter 6) is the **angiosperms** (*angios*, container, *sperma*, seed) or flowering plants. These plants produce seeds that are enclosed in and protected by the female reproductive structure of the flower called the **carpel**.

All seed plants are **heterosporous**, i.e., producing microspores and megaspores. Spores are produced inside structures called **sporangia** (singular, sporangium) which are commonly found associated with a reproductive leaf called a **sporophyll**. Details of the life cycle of Gymnosperms will be represented by the pine life cycle that is described later in this chapter. A general life cycle of heterosporous plants is shown in **Fig. 5.9**.

1) Become familiar with the general life cycle of heterosporous plants (**Figure 5.9**).

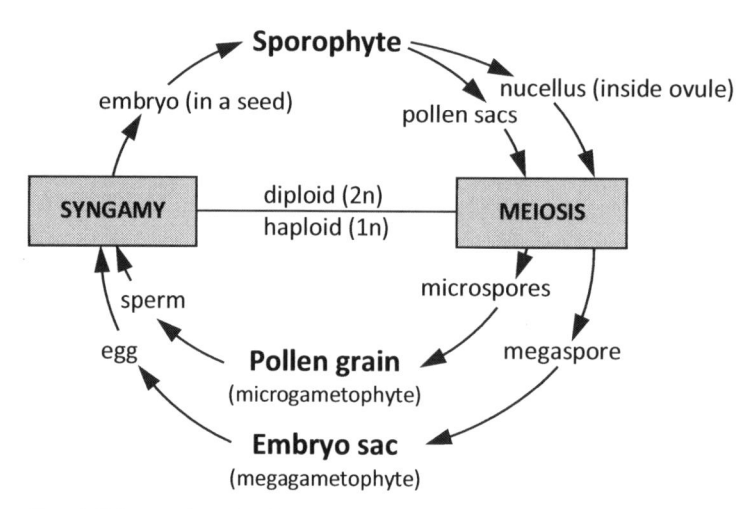

Figure 5.9. Generalized life cycle of a heterosporous (seed) plant. (Image: Dr. Gary Baird, Dept of Biology, BYU-Idaho)

In gymnosperms, the **microsporophylls** and **megasporophylls** are modified into scale-like structures which form **cones** or similar structures called **strobili**: **microstrobili**, male cones, and **megastrobili**, female cones, respectively. In male cones the **sporophylls** tend to be papery and short-lived. In female cones the sporophylls may be woody and persistent, and typically produce two **ovules** on the upper surface of each fertile scale. In most gymnosperms the female and male cones are produced on separate plants. Such plants are **dioecious**. In some conifers however, female and male cones are produced on the same plant, often female cones toward the top, and male cones father down. Such plants are **monoecious**.

There are four extant phyla of gymnosperms; all of which have an extensive fossil history. There are also many extinct spermatophyte taxa that are not described here. Each of the extant groups are introduced briefly.

Phylum Cycadophyta

The cycads are an ancient lineage of seed plants that were once quite diverse and worldwide in distribution. Today, there are about 140 cycad species, found primarily in southern Africa, Australia, Asia and the Caribbean. They are palm-like in appearance and produce pinnately (feather-patterned) compound leaves that are clustered at the top of a mostly unbranched trunk. Inside the thick trunk and large stems is a massive amount of starch. Some species have been harvested for their starch. Most cycads are extremely slow growing and a large cycad may be hundreds of years old. Many species are now endangered and the trade in cycad products is prohibited. Cycad seeds have been used for food but all parts of the plants are toxic, containing neurotoxins, hepatotoxins and carcinogens. All cycads form mutualistic associations with N-fixing cyanobacteria that are housed in their specialized roots. See **Figure 5.10** for an example of a living cycad.

Figure 5.10. King Sago cycad, a low-growing cycad. Note the pinnate fronds. Cones will be produced in the center of the crown of fronds when this plant is reproductive. (Image modified by ARH from a photograph by DanielCD, GNU Free Documentation license)

Phylum Ginkgophyta

Like cycads, the ginkgophytes were once diverse and widespread throughout the world. Today there is only one surviving species, *Ginkgo biloba*. It is often called a living fossil, and the fossil evidence of this species extents back millions of years. *Ginkgo* has long been grown in China, Korea and Japan, and is now cultivated around the globe. *Ginkgo* is a deciduous tree that can grow to 40 m tall and has fan-shaped leaves. Pollination is by wind and the resulting seeds are large. The mature female reproductive organs emit a strong, rancid odor (like feces) when stepped on or bruised. However, the inner part of the seed is edible and is a traditional Chinese food. See **Fig. 5.11** for an example of leaves, fruit and seed of the ginkgo tree.

Figure 5.11. Characteristic fan-shaped leaves of the gingko, two gingko fruits and seed. (Image modified by ARH from a photo courtesy of Koba-chan, GNU Free Documentation license)

Phylum Gnetophyta

The gnetophytes are a small group of about 70 species that are quite different from other gymnosperms and even from each other. They share a number of morphological, anatomical and developmental features in common with the angiosperms (flowering plants), most notably their conductive tissues. One representative of this phylum is shown in **Fig. 5.12**.

Phylum Pinophyta

This group, commonly called conifers, represent the largest group of extant gymnosperms with about 600 species worldwide, and includes pines, firs, cypresses and their relations. These are all woody plants and many of them are dominant features of our landscapes. In some regions, they form vast forests, as across northern Eurasia and North America. Contained within this group are the tallest known organism (coast redwood, 115.7 m/380 feet), the most massive individual organism known (sequoia, 1.2 million kg/2.7 million pounds) and oldest known living individual organism (bristlecone pine, 4,850 years old). Some conifers have great commercial value, being the source of most of our lumber and pulp for paper products.

Figure 5.12. The gnetophyte *Ephedra viridis,* commonly known as Mormon tea: male cones (left), bush (right). This is a semi-desert plant found throughout the Great Basin and American southwest. It can be used to brew a warm drink, but it is an acquired taste. (Images modified by ARH from photographs by By Dcrjsr - Own work, CC BY 3.0, https://commons.wikimedia.org/w/index.php?curid=16450961)

There are five extant families of conifers that are divided into two clades: pines and their relatives form one clade, and cypresses and their relatives form the other. Conifers may be either monoecious (mostly the pine clade) or dioecious (mostly the cypress clade). Female cones are most often woody. All conifers are wind pollinated and development of seeds can be a slow process. For example, it takes two years for the cones and seeds of pines to mature, while most other conifers require a single year. **Figure 5.13** shows the life cycle of a pine and is used here to represent conifer life cycles in general.

1) Become familiar with the pine life cycle (**Fig. 5.13**). A description of this life cycle is included below.
    a. The long-lived, mature diploid **sporophyte** stage (tree) dominates the life cycle. In most pines, sporophytes produce both female and male cones.
    b. Female cones are generally comprised of woody **sporophylls** (scales). Each scale bears two **ovules** and each ovule contains a **megasporangium**. A megasporangium produces four haploid **megaspores** by meiosis. One of these megaspores is larger than the rest and persists while the three smaller megaspores degrade and die.
    c. The surviving large megaspore undergoes mitosis and produces a tiny haploid **gametophyte** inside the megasporangium. Each **gametophyte** produces only two eggs, but only one of these will be fertilized. The scales on the female cone open when eggs are viable, exposing **micropyles** (openings) of **ovules** to the external environment so pollination can occur.
    d. In the meantime, male cones are also produced. Male cones are usually much smaller than female cones and their **sporophylls** (scales) are paper-thin. Male cones persist only long enough to produce and release **pollen grains**. Each scale of a male cone contains **microsporangia** where haploid **microspores** are produced by meiosis. Each **microspore** produces a **pollen grain** via mitosis. A pollen grain is a tiny male **gametophyte** that contains two nuclei.

One is the **pollen tube nucleus** and the other is the **sperm nucleus** that will be used during **syngamy**.

    e. When female and male cones are open, **pollen grains** are carried by the wind to female cones. **Pollination** occurs when a pollen granule lands on a cone, is drawn in through the **micropyle**, and contacts the exposed surface of a **megasporangium** – the structure that houses the **gametophyte** and eggs.

    f. Once a **pollen grain** contacts an ovule it produces a **pollen tube** that penetrates the megasporangium and carries a sperm nucleus to one of the eggs in the gametophyte that is housed there. **Syngamy** occurs when the sperm nucleus fuses with the egg nucleus and a **zygote** is formed. At this point, the scales of the female cone typically close again to protect developing embryos and to facilitate seed production.

    g. The zygote develops into an **embryo** that is protected by surrounding tissues and seed coat.

    h. When seeds are viable the female cones reopen and seeds are released. Depending on the species and environmental conditions, it can take days to years between seed release and germination.

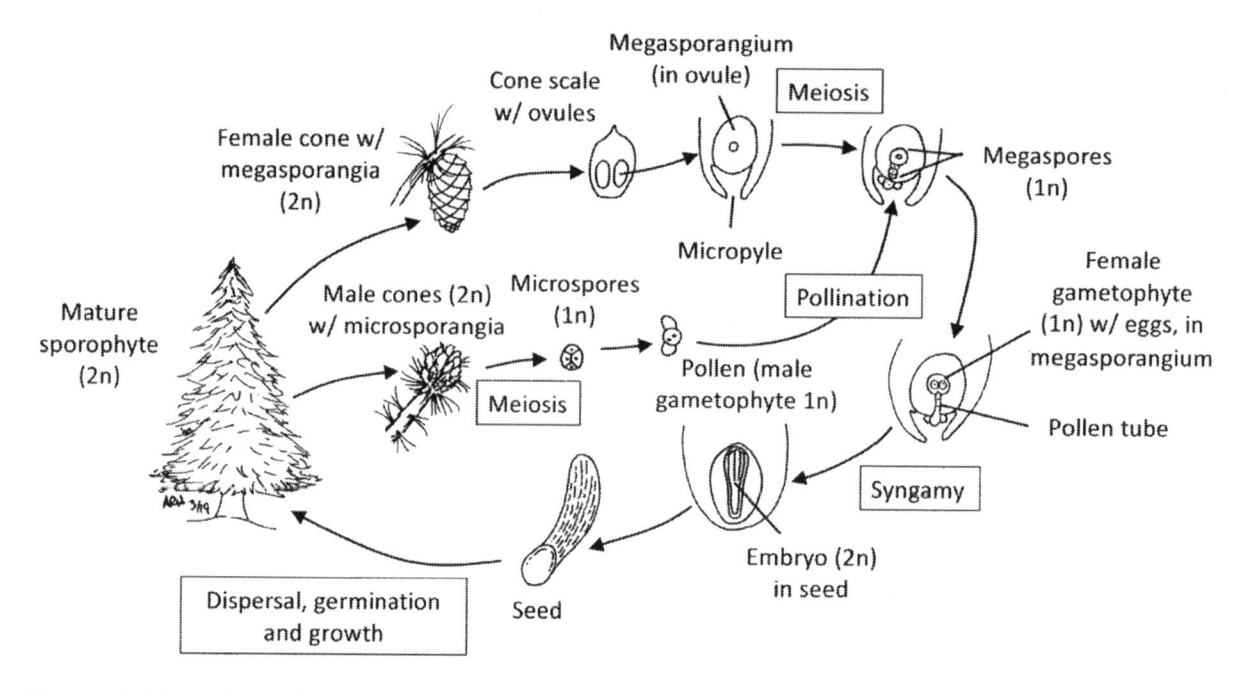

Figure 5.13. Life cycle of a pine. (Image: ARH)

2) Observe, describe and DRAW pine leaves (needles). Use both fresh material and prepared slides of *Pinus* leaf.

    a. Pine needles, macro-anatomy – Obtain some fresh pine needles. Observe the leaf (needle) arrangement. Note that the needles are arranged in **fascicles** (bundles). DRAW a bundle of needles. Next, make a cross-sectional cut through all needles of a fascicle and look at the cross-sectional shape of the needles as well as how each needle fits together with the other needles of its fascicle.

b. Pine needles, micro-anatomy – Obtain a prepared slide of a cross-section of a pine needle and examine it using a compound microscope. Observe, describe and DRAW what you see. The **dermal tissue** typically consists of the **epidermis**, the **hypodermis**, and **stomata**.

c. Note the **mesophyll cells** and the **endodermis** around the **vascular bundle**. Between the endodermis and the **xylem-phloem** is the **transfusion tissue**. This tissue aids in the movement of water and sugars. Imbedded among mesophyll cells leaf are **resin ducts or canals**. Be sure to label all the structures listed above. Use **Fig. 5.14** to help you identify what you see.

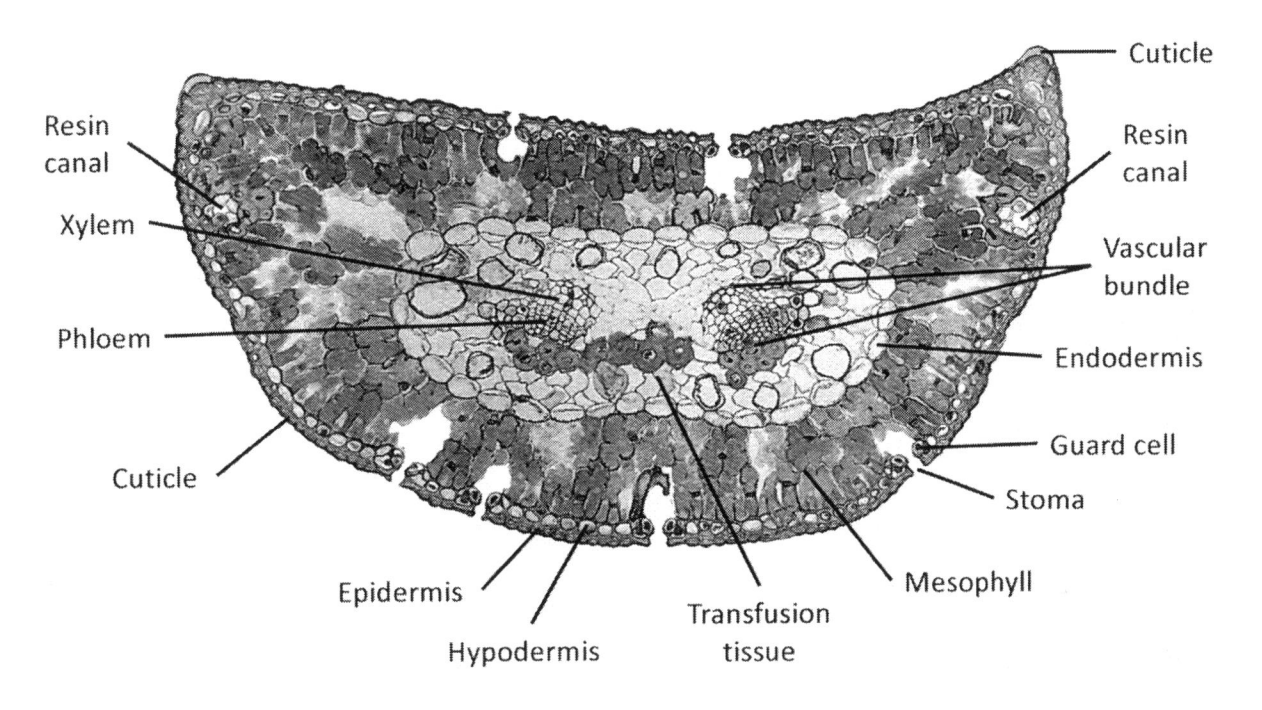

Figure 5.14. Cross section of a pine leaf (needle). (Image: ARH)

3) Cone anatomy – In conifers, female cones are deceptively complex. Each scale of the cone is actually a modified branch called a **seed-scale complex**. The seed-scale complex consists of a **sterile scale or bract** and a **fertile scale** that bears two **ovules**; hence this scale is called the **ovuliferous scale**. In pines, the exposed tip of a juvenile ovuliferous scale is called an **umbo**. As the cone matures, the tip of the ovuliferous scale enlarges and forms a structure called the **apophysis**. The umbo is still evident on the apophysis as a scar, prickle, hook, etc.

   a. Female cone – Observe and describe an unpollinated female pine cone (**megastrobilus**).

      i. Obtain a fresh female pine cone. Using a razor blade or scalpel, slice the cone in half length-wise and observe it with a dissecting scope or hand lens. On the cut surface, find a **seed-scale complex** that has been cut in half and observe what the seed-scale complex looks like from the side. DRAW a representation of the seed-scale complex from the side and label the following: **ovuliferous scale**, **sterile scale or bract**, and **ovule**.

ii. If fresh cones are not available or you are having difficulty locating a nicely sectioned seed-scale complex, use prepared slides of a female cone. To do so, use a microscope and observe a longitudinal section through an ovuliferous cone. Compare this to what you observed in the living material. DRAW and label the **ovuliferous scale**, **ovule**, **micropyle**, and **sterile scale or bract**. Do this ONLY if you have not already done this in the previous section. Refer to **Fig. 5.15** to help you identify what you see.

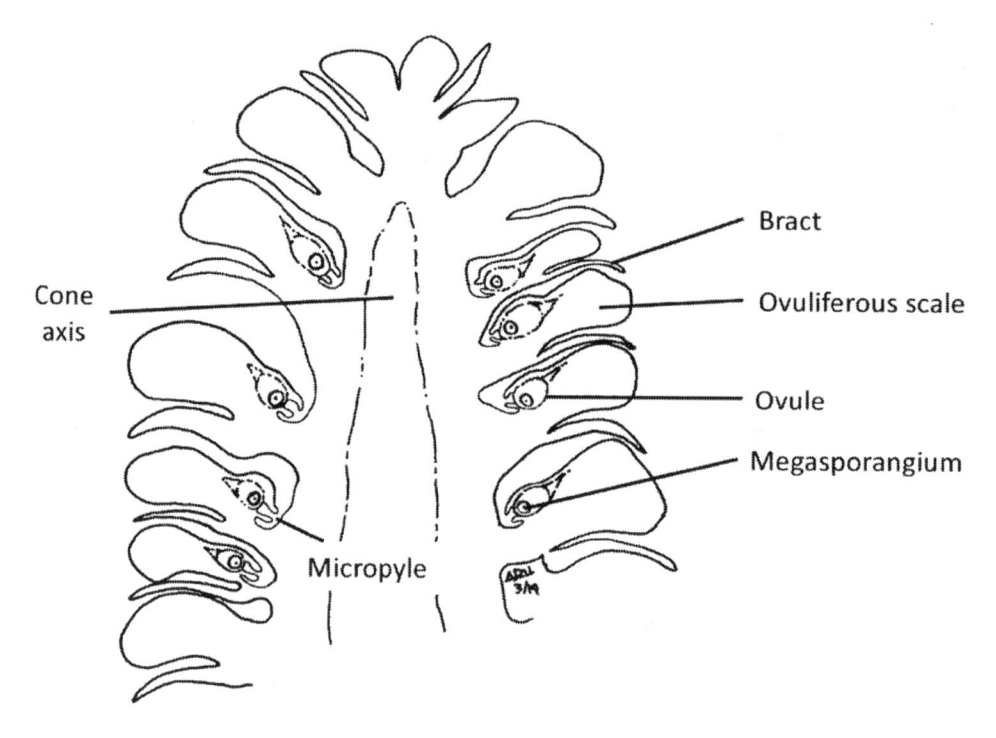

Figure 5.15. Cross-section view of an ovuliferous female pine cone. Note: Bracts are sterile scales. (Image: ARH)

b. Female cone – Using the same cone, locate an intact seed-scale complex and carefully remove it. Note the two **ovules** on the upper surface of the ovuliferous scale. The **micropyles** face the central axis of the cone. Draw a representation of an ovuliferous scale from the top and label the **ovuliferous scale** and **ovules**.

c. Observe examples of other ovuliferous cones from a variety of coniferous species, as assigned/available. Note the diversity of sizes, shapes, and architecture of the cones.

d. Observe a number of different mature cones. Select one cone and observe one of its seed-scale complexes. Locate the apophysis, umbo, ovuliferous scale, the two **seeds** or **seed scars** and the **sterile scale or bract** which will be small in pines. Draw upper and lower surfaces of a seed-scale complex for a fir, a spruce, and a Douglas fir. For each assigned species, label the **ovuliferous scale**, **sterile scale or bract** (if possible), and **seed** or **seed scar**.

Group Questions

1) If you review the brief introduction to the cycads, you will recall that their tissues produce various kinds of toxins. Cycads are not the only plants to contain poisonous compounds. Develop an evolutionary hypothesis that explains why plants may produce toxins, and explains where these toxins may have come from in the first place.
2) You may need to do a little research to answer this question, but develop a list of challenges organisms face when they live in on land. Next, develop a list of adaptations that allows seed plants like pines to overcome those ecological challenges.
3) Ferns tend to be found mainly in moist environments. Explain why this is so.
4) Develop a hypothesis that explains why female cones of conifers are so woody, and why male cones are not. Note that it's a large investment of energy to produce the heavy woody cones typical of conifers.

Non-flowering vascular plants – Glossary

Angiosperms – flowering plants

Annulus – single layer of cells surrounding most of a leptosporangium, aids in spore release

Antheridium/antheridia – structure(s) that produces sperm

Apical meristem – region of growth and elongation in growing tips of stems, branches and root

Apophysis – swollen tip of a mature ovuliferous scale

Archegonium/archegonia – structure(s) that produces eggs

Bract – see sterile scale

Capsule – structure that houses spores in leptosporangiate ferns

Carpel – female structures of a flowering plant

Circinnate vernation – the way a new fern frond emerges, unrolling to protect the young frond

Cone axis – central structural support of a cone

Cone scale – structure that bears ovules

Cones – see microstrobili and megastrobili

Cuticle – acellular, water-resistant outer covering of plants

Dermal tissue – tissue covering the outer surface of a body

Dioecious – separate sexes

Embryo sac – structure that produces an egg in a megasporangium

Enation – leaf-like outgrowth that lacks true vascular tissues

Endodermis – layer of cells that surrounds vascular tissue in roots and some stems

Epidermis – outermost layer of cells of the dermis

Euphylls – leaves that have true vascular tissues

Fascicle – a bundle of needles (in pines)

Fertile scale – see ovuliferous scale

Fiddlehead – newly forming fern frond

Fronds – leaves of ferns

Gametophyte – life stage that produces gametes

Guard cells – cells that line and regulate the size of openings of stomata

Gymnosperms – plants that have seeds that directly exposed to the environment

Heterosporous – plants that produce spores that are different sizes

Homosporous – plants that produce spores that are all the same size

Hypodermis – layer of cells underlying the epidermis

Indusium – thin flap of tissue that protects developing fern sporangia

Leptosporangium – a sporangium that develops from a single cell, has a wall one-cell layer thick that includes an annulus that aids in spore release, and produces relatively few spores

Lignin – complex compound found in cell walls of rigid or woody tissues of plants

Marginal meristem – growth regions on opposite sides of a stem that produce leaves as well as leaf shapes, increases width and number of cell layers

Megasporangium – structure that produces large spores (megaspores)

Megaspore – a haploid cell produced by meiosis that develops into the female gametophyte

Megasporophylls – leaves that produce megaspores

Megastrobili – female cones that are woody and persistent

Mesophyll – inner tissue of leaves that contain lots of chloroplasts

Micropyle – opening into an ovule through which pollen can enter

Microsporangium – structure that produces small spores (microspores)

Microspore – spore that will produce pollen grains

Microsporophylls – leaves that produce microspores

Microstrobili – male cones with paper-like scales, and are short-lived

Monoecious – hermaphroditic, produces both eggs and sperm

Nucellus – structures inside the ovule

Ovule – structure in a female cone that houses a megasporangium

Ovuliferous scale – scale that bears ovules

Phloem – vascular tissue that conducts (transports) photosynthate (sugars)

Pollen grain/granule – microgametophyte of seed plants

Pollen tube – structure produced by a pollen grain that carries the sperm nucleus to an egg inside of a female gametophyte

Prothallium – heart-shaped haploid gametophyte stage of ferns

Resin ducts/canals – ducts through which resin is transported

Rhizoid – root-like filaments

Rhizome – horizontally growing root system that usually supports multiple vertical shoots

Seed – structure that houses an embryo and nutritive material, protected by a tough seed coat

Seed-scale complex – the sterile scale or bract and a fertile or ovuliferous scale

Seed scar – mark left on an ovuliferous scale after the seed is released

Sorus/sori – cluster(s) of leptosporangia on the underside of a fern frond

Sperm nucleus – nucleus in a pollen grain that will fertilize an egg

Spore – single celled life stage that can germinate and grow without first undergoing syngamy

Spermatophyte – plants that produce seeds

Sporangia – structures that produce spores

Sporophyll – leaf that produces spores

Sporophyte – life stage that produces spores

Stele – central core of tissue in a stem that includes vascular tissues and supporting tissues

Sterile scale/bract – non-reproductive scale that is paired with an ovuliferous scale in cones

Stomata – opening in the dermis allowing gas exchange, the size of the opening is regulated by guard cells

Stomum – cells where a leptosporangium ruptures when annulus cells contract

Strobilus/strobili – alternative term for cone(s)

Syngamy – fusion of egg and sperm nuclei and creation of a zygote

Thallophyta – non-vascular plants (see Ch 4)

Tracheid – specialized cell with thickened cell walls found in xylem

Tracheophyta – plants that have xylem and phloem

Transfusion tissue – tissue that facilitates movement of materials in vascular tissue

Umbo – tip of immature ovuliferous scale

Vascular bundle – cluster of xylem and phloem cells

Xylem – vascular tissue that conducts (transports) water and dissolved minerals

Zygote – product of fusion of egg and sperm nuclei

Chapter 6: Plantae - Flowering Vascular Plants

The flowering plants, also known as **Angiosperms** (*angeion*, vessel, *sperma*, seed), differ from the non-flowering seed plants in Chapter 5, because angiosperms produce flowers and their seeds are housed inside an ovary. Angiosperms are by far the most biologically diverse and ecologically successful group of plants on Earth today. There are over 300,000 described species of angiosperms worldwide, occurring in almost every conceivable terrestrial environment as well as freshwater and marine habitats. The presence of flowers is such a unique feature that angiosperms are also referred to as the **Anthophyta** (*anthos*, flower, *phyton*, plant). All flowering plants are members of the monophyletic **Phylum Magnoliophyta.**

Phylum Magnoliophyta has traditionally included two classes: Magnoliopsida or Dicotyledonae (**dicots/eudicots**), and Liliopsida or Monocotyledonae (**monocots**). This arrangement has been challenged by several lines of evidence. Phylogenetic analysis shows that the monocots are a monophyletic lineage, but the dicots are not. Instead, there are a number of small basal taxa of angiosperms that represent ancestral lineages that do not fit into either the monocot or eudicot clades. Monocots and eudicots diverged from each other after these basal lineages evolved. Today there are approximately 60,000 species of monocots with most of these belonging to a few species-rich groups (e.g., orchids, lilies, grasses, sedges, and palms). The eudicots contain perhaps 250,000 species, and is the largest group of angiosperms (**Fig. 6.1**).

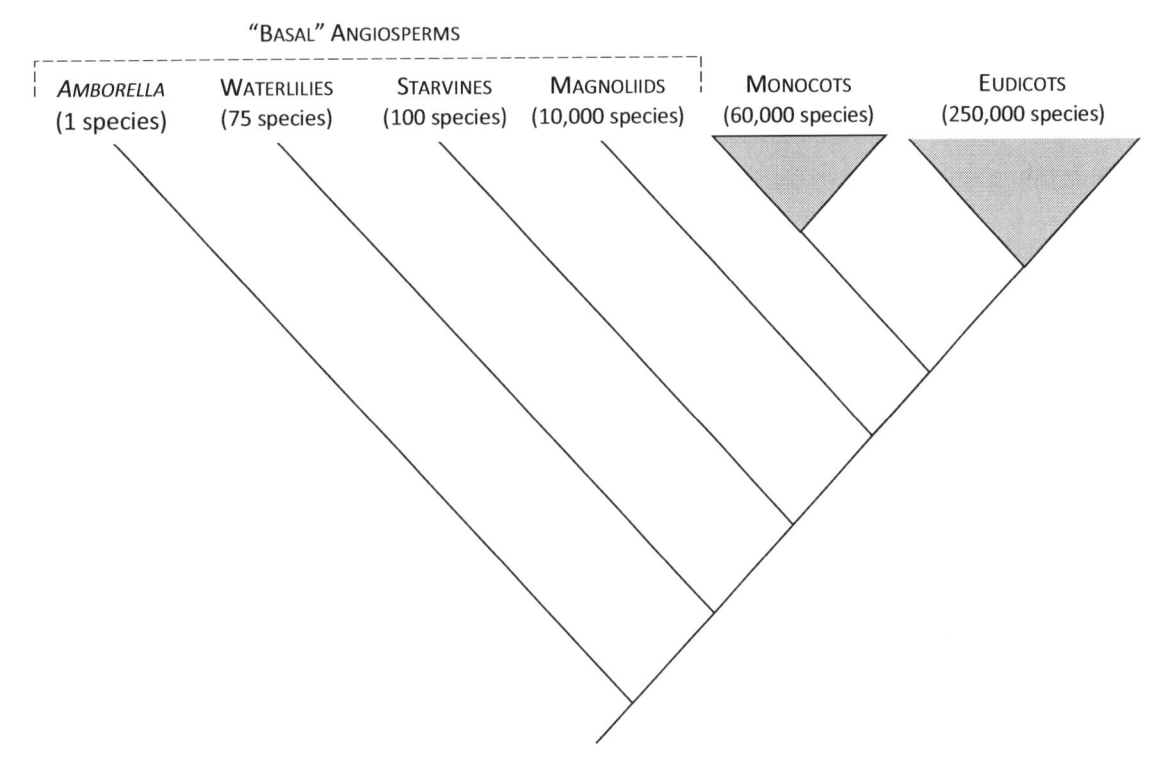

Figure 6.1. A phylogenetic tree of angiosperms.

Morphological and anatomical characteristics distinguish monocots from eudicots. Not all of these features are completely unique to one or the other, but using the entire suite of features can provide a way to differentiate between the two groups (see **Table 6.1**).

Table 6.1. Comparison of characteristics of monocots and eudicots.

Characteristic	Eudicots (dicots)	Monocots
Stem vascular tissue	Bundles in a ring	Bundles scattered
Vascular cambium	Commonly present	Absent or rare
Root system	Tap rooted	Fibrous rooted
Leaf venation	Net veined	Parallel veined
Floral parts	Mostly in sets of 4 or 5	Mostly in sets of 3
Cotyledons ("seed leaves")	Two	One
Pollen grains	Triaperturate (3-pores)	Monoaperturate (1-pore)

The evolution of **flowers** was another important evolutionary step. Flowers are shoots that produce both sterile and fertile modified leaves, and are used to facilitate reproduction. The structure of all flowers follows the same basic pattern, including four whorls or rings of structures. The outermost two whorls of a flower are sterile and comprise the **perianth**. The outer of these whorls is the **calyx** and is made of **sepals**. These are usually green and enclose and protect the flower during the bud stage. The inner of these two whorls is the **corolla** and is made of individual parts called **petals** which may be brightly colored. Flowers use this bright coloration to help attract animal pollinators. When sepals and petals are indistinguishable, then the term **tepals** is applied to all perianth parts. The inner two whorls of a flower are modified fertile leaves. Male organs of a flower form the first whorl inside the perianth and are collectively referred to as the **androecium**. Each male organ is a **microsporophyll** called a **stamen**. Each stamen includes a **filament** and an **anther**. The innermost whorl is the female part of a flower, and is referred to as the **gynoecium**. The female organ is called a **pistil**. It consists of an **ovary** that contains immature **ovules** and a stalk-like **style** with a tip called the **stigma**, where pollen grains land.

Though all flowers are based on the same basic plan, angiosperms exhibit an astonishing degree of variation in floral morphology, largely due to the pollination mechanism employed. Most pollinators are insects, but there are also bird and bat pollinators. There are also many plants that are wind-pollinated.

Although the order of floral whorls does not vary, the presence, number and symmetry of floral parts varies greatly. A **complete flower** is one that produces all four whorls. An **incomplete flower** is one in which one or more of the whorls is missing, but the remaining whorls are still in the correct order. A **perfect flower** contains both an androecium and gynoecium, whereas an **imperfect flower** is missing one or the other of these two whorls. Flowers that produce only an androecium are called **staminate flowers**, and those that produce only a gynoecium are called **pistillate flowers**. If a plant produces both staminate and pistillate flowers, it is **monoecious**. If a plant produces only staminate or only pistillate flowers, it is **dioecious**.

Task #1: Flower anatomy

1) Flower anatomy – obtain a fresh flower or study a model of a flower. Observe, describe and DRAW your flower. Label the four whorls of the flower (outer to inner – **sepals**, **petals**, **stamens** and **pistil**). Refer to **Fig. 6.2** to help you identify these structures. Also, indicate its symmetry (radial, biradial or asymmetrical) and whether it is complete or incomplete. If it is not complete, indicate the missing structure(s).

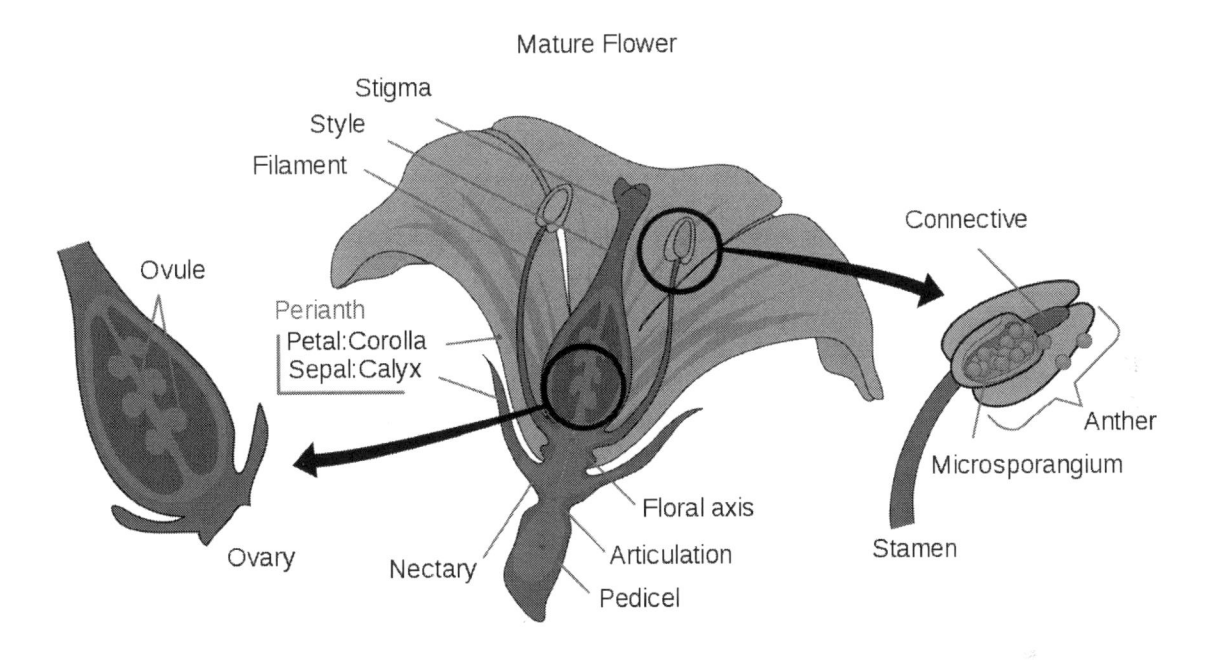

Figure 6.2. Anatomy of a perfect, complete flower. (ARH modified image, public domain by Mariana Ruiz – LadyofHats, Wikimedia Commons)

2) Either during lab, or some other time this week, take a walk outside if flowers are blooming or visit a greenhouse and look for different kinds of flowers. Observe them in place (do not pick them). You can use an APP like PlantSnap to identify the kinds of plants you find. Produce a table in your notebook that includes the traits listed below and do the best you can to fill it out for each kind of flower you observe.
   a. What is the flower symmetry?
   b. Is the flower complete? If not, indicate the missing structure(s).
   c. Is the flower perfect? If not, which structures are missing?
   d. How many sepals and petals are present?
   e. Is it a monocot or eudicot plant?
   f. How many stamens are in the androecium?
   g. How many pistils are in the gynoecium?

Task #2: Reproductive structures of the flower

1) Male reproductive structures – Expose the inner whorls (reproductive structures) of your flower by carefully removing all sepals and petals from one side of your flower. The whorl of structures just inside the petals is the **androecium**.
   a. Observe, describe and DRAW one of the **stamens** (**anther** and **filament**). Refer to **Figs. 6.2** and **6.3** to help you identify these structures.
   b. Obtain a prepared cross section slide of an anther. Use a compound microscope to observe, describe and DRAW what you see. Refer to **Fig. 6.3** to help you identify what you see. Label the structures of the pollen sac as indicated on **Fig. 6.3**.

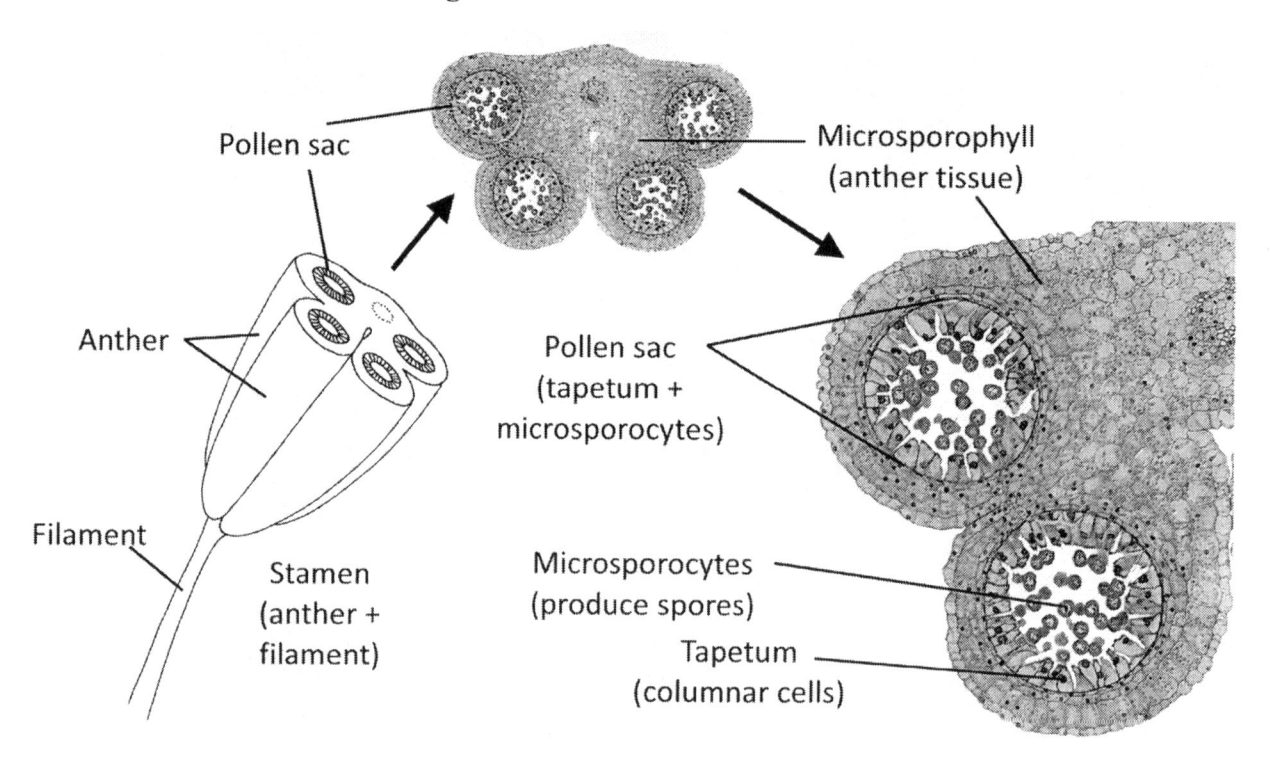

Figure 6.3. Androecium of a flower. **Left:** Cross-section through a bilobed anther, showing the two pollen sacs per lobe. **Upper center:** Photomicrograph of cross section through the anther. **Right:** Photomicrograph of two pollen sacs of one lobe of an anther. (Images: ARH)

2) Female reproductive structures – The female organ(s) of a flower comprise the innermost whorl, and are referred to as the **gynoecium**.
   a. Observe, describe and DRAW the external anatomy of the female structures of the flower from which you removed the **perianth**. Be sure to label the **pistil** including the **style**, **stigma** and **ovary**. Refer to **Fig. 6.2** to help you identify these structures.
   b. Anatomy of the pistil – You will need two flowers for this part of the exercise, or you can team up with a neighbor or another lab team. One student or team should do the longitudinal dissection and the other student or team should do the cross-section dissection, then show each other what you see.

i. Longitudinal section of the pistil – use a razor blade to make a longitudinal cut through the entire **pistil** (**stigma**, **style** and **ovary**). Observe, describe and DRAW what you see. Refer to **Fig. 6.4** – right image, to help you identify these structures.
ii. Cross-section of the ovary – use a razor blade to make a cross-section cut through the ovary. Observe, describe and DRAW what you see. Refer to **Fig. 6.4** – left image, to help you identify structures.
iii. Discuss with a neighbor or lab partner what the three-dimensional structure of the ovary looks like. Record these ideas in your lab notebook.

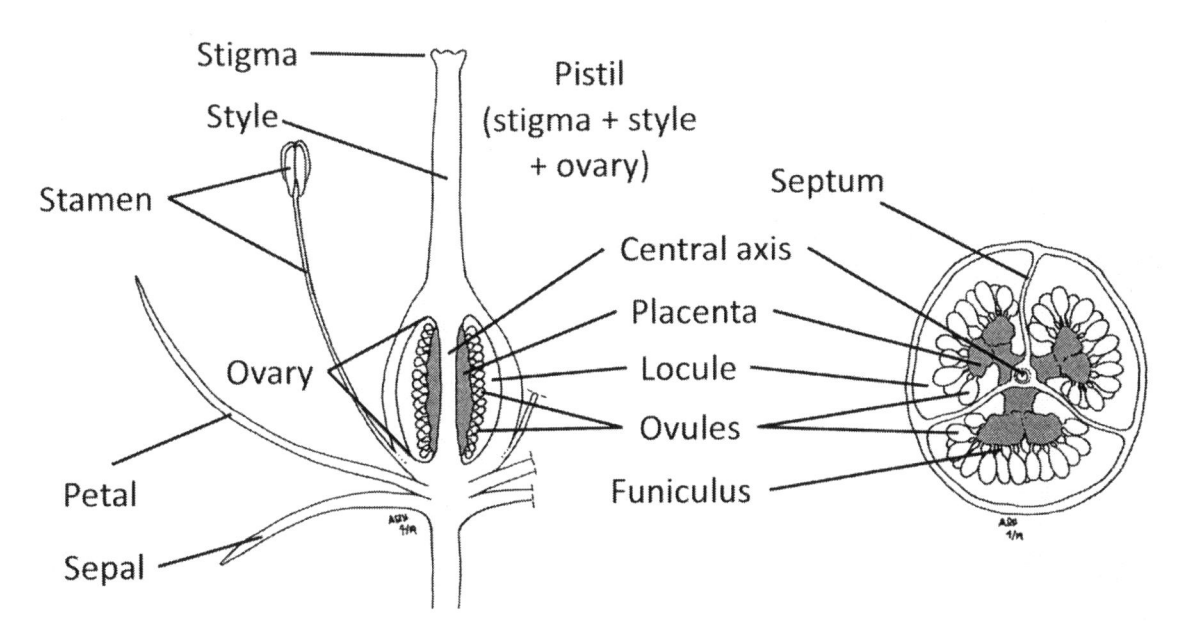

Figure 6.4. Gynoecium (female reproductive parts) of a flower. **Left:** Longitudinal section of a pistil. **Right:** Cross-sectional through an ovary. (Images: ARH)

Task #3: Angiosperm life cycle and reproduction

Become familiar with the life cycle of flowering plants. Refer to **Fig. 6.5** as you work through the following descriptive information of the angiosperm life cycle.

1) Pollen production
   a. Inside each two-lobed **anther** are two pairs of **microsporangia** called **pollen sacs**. The development of **pollen grains** is similar to that observed in other **spermatophytes** but the **microgametophyte** is even more reduced.
   b. Within each pollen sac is an outer layer of columnar cells called the **tapetum**, and an inner layer of **sporogenous tissue**. The sporogenous tissue is composed of diploid **microsporocytes** that undergo meiosis to produce four haploid **microspores** each.

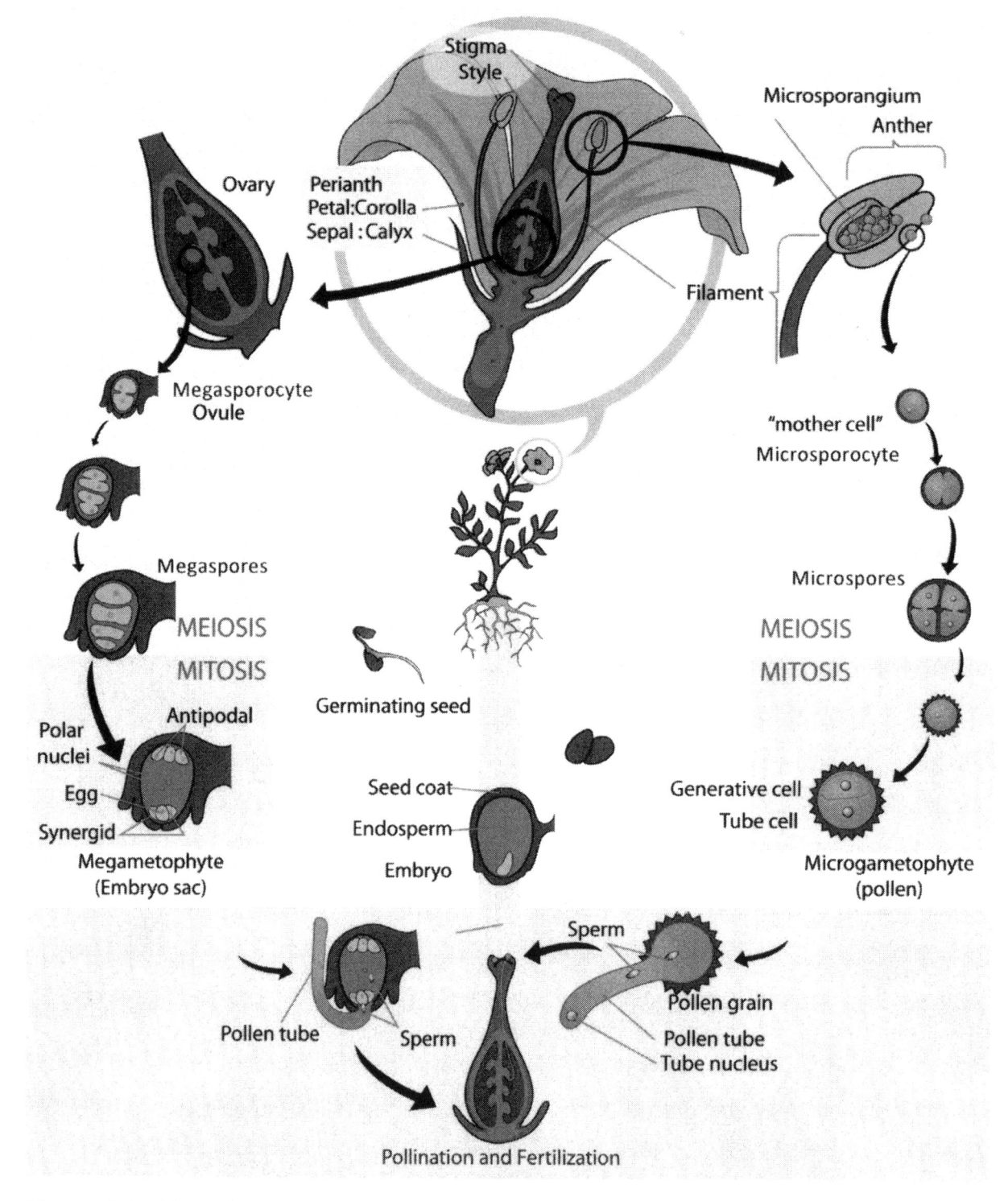

Figure 6.5. Life cycle of a flowering plant. (Image modified by ARH from a figure in the public domain produced by Mariana Ruiz – LadyofHats, Wikimedia Commons)

    c. Each microspore develops into a pollen grain with a thick, two-layered cell wall. The inner wall or **intine** is composed of cellulose and hemicelluloses and is generated by the microspore. The outer wall or **exine** is composed of

sporopollenin, an environmentally resistant biopolymer that is produced by the tapetum. Each microspore divides mitotically to produce a larger **tube cell** and smaller **generative cell**. The generative cell is completely surrounded by the tube cell. The generative cell later divides to produce two **sperm cells**. When this three-celled stage is achieved the pollen grain (microgametophyte) is mature.

2) Egg production
    a. The **nucellus** is tissue inside an ovule that houses a single cell called the megasporocyte which undergoes meiosis to produce four haploid **megaspores**.
    b. The three megaspores closest to the **micropyle** disintegrate, leaving a single megaspore. This megaspore divides mitotically three times without cytokinesis to produce an immature **embryo sac** composed of a single **coenocytic cell** with eight nuclei.
    c. Three of these nuclei then become **antipodal cells** located at the end of the embryo sac opposite the micropyle. Three other nuclei are formed into the **egg** and two **synergid cells** which are located next to the micropyle. The last two nuclei, called **polar nuclei**, remain within the large **central cell**. When this 7-celled stage has been achieved, the embryo sac is mature.

3) Pollination
    a. Pollination occurs when pollen grains land or are deposited on the stigma of a flower, either by wind or an animal pollinator.
    b. If there is a same-species match between stigma and pollen grain, the **tube cell** of the pollen grain elongates and pushes its way through the **style** down into the **ovary** until it reaches the micropyle of an **ovule**.
    c. The elongating the **tube cell** carries the two sperm cells with it and releases them through the **micropyle** into the **ovule**.

4) Double-fertilization
    a. Two fertilization events happen after the sperm cells reach the embryo sac
        i. One fertilization event occurs when one of the sperm cells from the pollen grain fuses with the egg cell and **syngamy** occurs, forming a diploid **zygote** that will develop into the **embryo**.
        ii. The other fertilization event occurs when the second **sperm cell** fuses with two **polar nuclei** of the **central cell**, forming a triploid (3n) **endosperm cell**. The endosperm cell produces **endosperm tissue**. Endosperm is the nutritive tissue that will support the developing embryo.

5) Fruit production
    a. After eggs in the ovules are fertilized, each ovule develops into a mature **seed** with a **seed coat** derived from **integuments** of the ovule.
    b. While seeds are being formed, the entire ovary becomes a fruit with the mature seeds protected inside. Seeds housed inside of fruits is a distinguishing characteristic of angiosperms.

6) Germination

a. When a seed receives the appropriate environmental stimulus, the **embryo** uses nutrients in the endosperm to support it during early stages of growth and development.
b. During germination, the seed coat splits and the embryo emerges and grows into a juvenile sporophyte.

Task #4: Fruits and seeds

Many organisms, including humans, are highly dependent on fruits for food, e.g., apples, oranges, tomatoes, beans, rice, wheat, corn, nuts, etc. And only angiosperms produce true fruits.

Angiosperm female reproductive organs are derived from fertile leaves (**megasporophylls**) called **carpels**. Evolutionarily, ovules would originally have been formed on exposed surfaces of these leaves, but the edges of carpels in angiosperms evolved to become sealed together to create internal spaces where **ovules** were isolated and protected from the external environment. One or more carpels can be fused together to collectively form the **ovary** of a flower. Generally speaking, the number of **locules** (seed chambers) inside a fruit corresponds to the number of carpels that made up the flower. This is actually where flowering plants get their name – angiosperms (*angeion*, container, and *sperma*, seed). Carpels also evolved a **stigma** to receive pollen and a **style** to facilitate sexual reproduction, thereby forming the **pistil** of a flower.

The purpose of the fruit is to aid in protecting and dispersing seeds. Seeds in mature fruit may be dispersed by wind, water, attached to the coats of animals, or even in the digestive tracts of animals that ate fruit. In the latter case, non-digested seeds are released with the feces.

The ovary often undergoes significant changes as a fruit develops. At fruit maturity, the ovary wall is called the **pericarp**. Classification of fruit types is often based on features of the pericarp. In some fruits, mostly those that are fleshy, the pericarp may consist of distinct layers: the **exocarp** – outer layer or skin, **mesocarp** – the middle, fleshy layer, and **endocarp** – the inner, sometimes hardened layer, as in the hard, outer layer of the pit of a cherry. The tomato you will observe has an exocarp and mesocarp, but no endocarp.

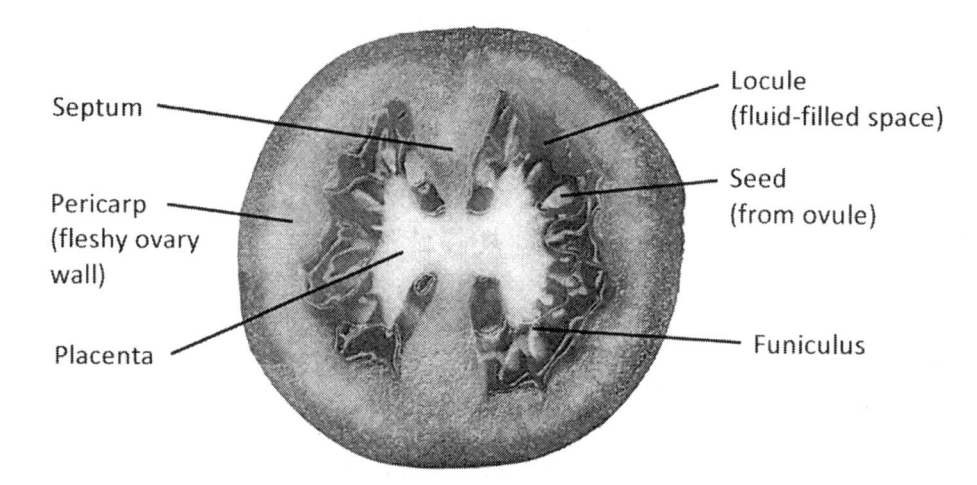

Figure 6.6. Anatomy of the fruit of a tomato. (Image modified by ARH from a photograph provided by Dr. Gary Baird, BYU-Idaho)

1) Become familiar with fruit anatomy
    a. Obtain a tomato and make a cross-section cut through it.
    b. Observe, describe and DRAW what you see. Refer to **Fig. 6.6** to help you identify fruit anatomy.
2) Become familiar with seed anatomy.
    a. Obtain a bean seed. A seed is a matured ovule and contains three basic structures: **seed coat**, **embryo**, and nutritive tissue. The seed coat is derived from the **integuments** of the ovule.
    b. Use a dissecting microscope or hand lens as you observe, describe and DRAW the bean seed. Label the the scar-like structure called the **hilum** where the **funiculus** was attached to the **ovule**. Adjacent to the hilum is a small opening. This was the **micropyle** of the ovule. Include this in your drawing. Refer to **Fig. 6.7** to help you identify these structures.
    c. Remove the seed coat from your bean seed. Observe, describe and DRAW what you see. The bean is a eudicot so all of the nutrition for the **embryo** is contained within the two large **cotyledons** – the large two halves of the seed. Separate the two cotyledons and observe the embryo inside. Identify the **radicle** – embryonic root, and the **plumule** – embryonic shoot. The radicle is the pointed end of the embryo and the plumule will be the two, small, leaf-like structures. Refer to **Fig. 6.8** to help you identify the internal structures of the bean seed.

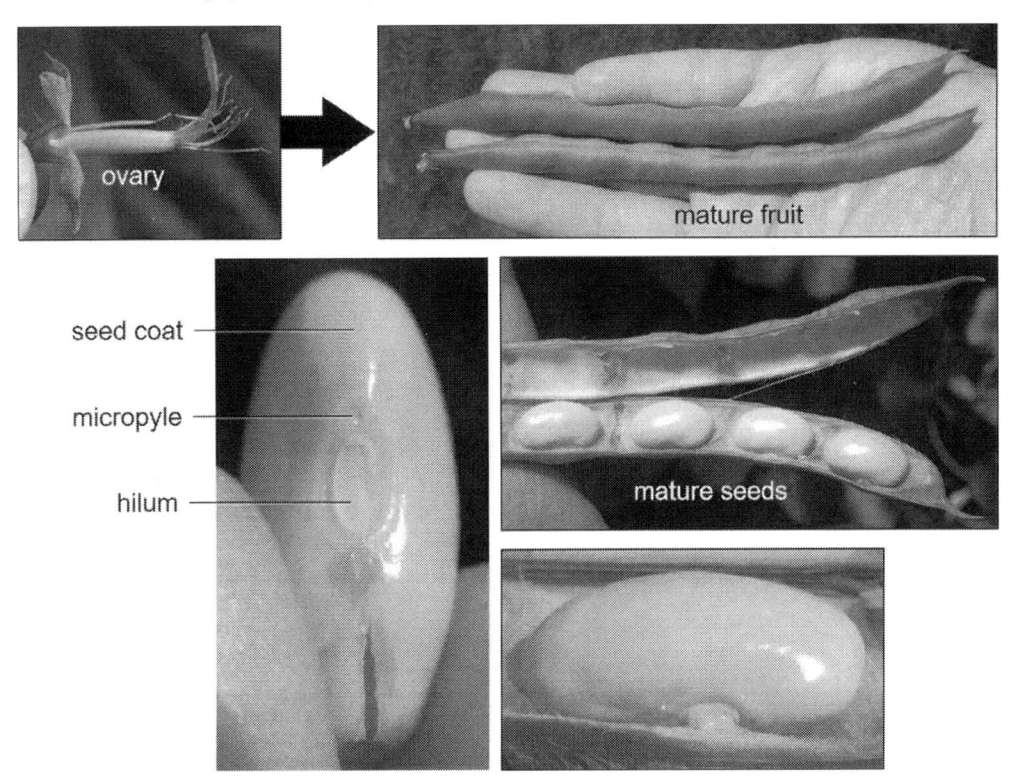

Figure 6.7. Top left: A bean flower and ovary. **Top right:** mature fruit. **Middle right:** mature seeds in the fruit. **Bottom right and left:** a seed with funiculus, and external anatomy of a bean seed (bottom left). (Image modified by ARH of a figure provided courtesy of Gary Baird, BYU-Idaho)

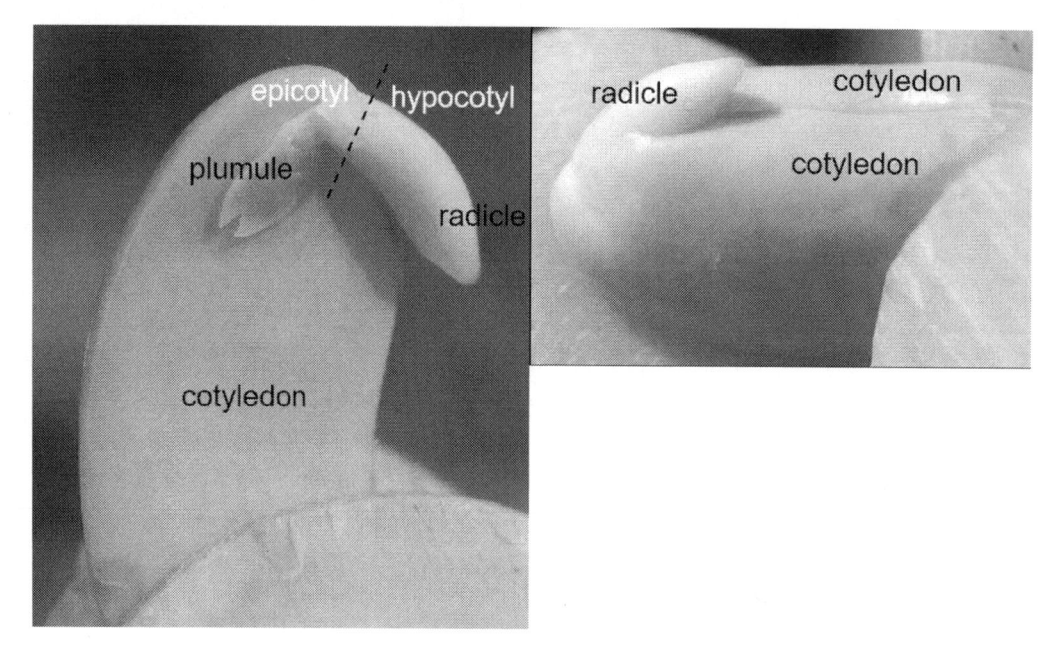

Figure 6.8. Anatomy of a bean seed. (Image modified by ARH of a figure provided by Gary Baird, BYU-Idaho)

Group Questions

1) Compare the energetic costs and benefits of animal-pollinated plants and wind-pollinated plants.
2) You may need to do a little research for this question, but explain how plants with perfect flowers avoid self-fertilization.
3) It is extremely costly in terms of energy for plants to produce large, fleshy fruits that are loaded with sugar, e.g., apples. Develop an evolutionary hypothesis that explains the history behind why plants started producing these kinds of fruits in the first place, and why they continue to do so.

Flowering Plants – Glossary

Androecium – first whorl of structures inside the perianth, male reproductive organs

Angiosperms – taxon containing all flowering plants, see also Anthophyta

Anther – microsporophyll that produces microspores and then pollen grains

Anthophyta – taxon containing all flowering plants, see also Angiosperms

Antipodal cells – haploid daughter cells of the surviving megaspore, and are located in the ovule on the wall opposite the micropyle

Bilateral symmetry – structure that can be divided into two mirror images in only one plane of symmetry

Calyx – outermost whorl of structures of a flower, made up of sepals

Carpel – a megasporophyll leaf in the innermost whorl of a flower that folds up on itself to produce an internal space (locule) where ovules are produced, and later produces a style and stigma. Generally, the number of locules in an ovary indicates the number of carpels that contributed to it.

Central axis of a pistil – pillar of tissue running down the center of an ovary from top to bottom

Central cell – large cell in the center of the embryo sac

Complete flower – flower that has all four whorls of structures

Corolla – inner whorl of the perianth, is comprised of petals

Cotyledon – nutrient-rich structure in a seed

Dicot – see eudicot

Dioecious – plants that produce either female or male flowers

Egg cell – daughter cell of the surviving megaspore that is located at the micropyle of the ovule and will be used to produce a zygote

Embryo – immature individual protected inside the seed coat

Embryo sac – structure in the ovule comprised of antipodal cells, the egg cell, synergid cells, and central cell

Endocarp – innermost layer of the pericarp, may be soft tissue or extremely hard, like the covering of a pit in a cherry, or not present as in tomatoes

Endosperm – nutritive material found in some seeds, produced by the endosperm cell

Endosperm cell – cell resulting from triple fusion of a sperm nucleus and two polar nuclei

Epicotyl – portion of the plant embryo that includes the plumule

Eudicot – flowering plants that produce seeds with two cotyledons, see also Table 6.1

Exine – outer wall of a pollen grain, made of environmentally resistant sporopollenin

Exocarp – outer layer or skin of a fruit

Filament – stem-like structure that supports the anther

Flower – shoots that produce structures that facilitate reproduction

Fruit – complex structure derived from the ovary, houses seeds in internal spaces called locules

Funiculus – narrow neck of tissue that connects ovules and later seeds to the placenta in the ovary

Generative cell – small cell in a pollen grain that is completely enclosed within the tube cell, and will produce two sperm cells

Gynoecium – innermost whorl of structures in a flower, female reproductive organs

Hilum – scar-like structure where a seed was attached to the placenta by its funiculus

Hypocotyl – portion of the plant embryo including the radicle

Imperfect flower – flower missing either the androecium or gynoecium

Incomplete flower – flower missing one or more of the four basic whorls of structures

Integuments – outer layers of the ovule that become the seed coat

Intine – inner wall of a pollen grain, made of cellulose and hemicelluloses

Locule – space inside an ovary that houses ovules

Megagametophyte – see embryo sac

Megaspore – spore produced in the ovule by meiosis of a megasporocyte

Megasporocyte – cell in an ovule that undergoes meiosis to produce megaspores

Megasporophyll – fertile leaf of the gynoecium that gives rise to a carpel

Mesocarp – middle, sometimes fleshy layer of a fruit, e.g., the white flesh of an apple

Microgametophyte – structure that houses the microsporangium, pollen grains in flowering plants

Micropyle – opening into the embryo sac

Microsporangium – structure that will produce microspores, the anther in flowering plants

Microspore – haploid cells that develop into the male gametophyte

Microsporocyte – small cells in the middle of a pollen sac that produce microspores by meiosis

Microsporophyll – structure that produces microspores and then pollen grains

Monocot – flowering plants that produce seeds with one cotyledon, see also Table 6.1

Monoecious – organism that produces male and female flowers

Nucellus – tissue in an ovule that houses the megasporocyte

Ovary – female structure at the base of the pistil, produces ovules

Ovule – structure that produces megaspores, then the female gametophyte and an egg

Perfect flower – has androecium and gynoecium

Perianth – outer two whorls of a plant, made up of the calyx and corolla (sepals and petals)

Pericarp – the ovary wall that becomes the part of the fruit that surrounds the seed-containing locule(s), and may have up to three distinctive layers – the exocarp, mesocarp and endocarp

Petal – inner whorl of sterile leaves of the perianth, are often brightly colored to attract pollinators

Pistil – female organ, includes the ovary, style and stigma, and is the product of one to many carpels

Pistillate flower – flower missing the androecium

Placenta – soft tissue that initially supports ovules and later, seeds in an ovary

Plumule – embryonic shoot

Polar nuclei – two nuclei in the central cell of the embryo sac

Pollen grain – male gametophyte, houses the pollen tube nucleus and sperm nucleus

Pollen sac – space inside an anther where microspores and pollen grains are produced

Pollen tube – elongate structure produced by the tube cell of a pollen grain that carries sperm cells to an ovule housing an egg

Pollination – when a pollen grain of the correct species lands on the stigma

Pollinator – organism that facilitates pollination by transporting pollen grains from one flower to another

Radial symmetry – structure that can be divided into two mirror halves in multiple planes of symmetry

Radicle – embryonic root

Seed – structure that has a resistant seed coat, nutritive material and plant embryo

Seed coat – environmentally resistant, sometimes poisonous outer covering of a seed, derived from the integument of the ovule

Sepal – sterile leaves that comprise the calyx of a flower

Septum – tissue wall dividing locules from one another in an ovary

Sperm nuclei – nuclei produced by the generative cell of a pollen grain, one fertilizes the egg and the other contributes to nutritive tissue of the seed

Sporogenous tissue – made up of microsporocytes in the middle of a pollen sac, and produce microspores

Sporopollenin – environmentally resistant polymer produced by the tapetum, and form the outer layer of pollen grains

Stamen – male reproductive microsporophyll, anther and filament

Staminate flower – flower missing the gynoecium

Stigma – tip of the style, where pollen grains land during pollination

Style – structure of the pistil that bears the stigma

Synergid cells – daughter cells of the surviving megaspore that flank the egg cell

Syngamy – fusion of egg nucleus and sperm nucleus to produce a zygote

Tapetum – columnar cells in a pollen sac that produce sporopollenin

Tepal – term use to refer to sepals and petals when they cannot be differentiated

Tube cell – larger cell of a pollen grain that houses the generative cell

Zygote – cell produced when the nucleus of an egg and sperm fuse

Chapter 7: Fungi

Fungi is a diverse taxon that occurs in almost all habitats. There are about 120,000 known species of fungi but this may be only a small portion of the true diversity of this group. Estimates place the number of fungus species as high as 5,000,000; we simply don't know. Most fungi are **saprobes**, they are heterotrophic with absorptive digestion. They secret digestive enzymes that allow them to break-down just about any organic substance, natural or man-made. This makes fungi highly efficient decomposers, and their ecological significance cannot be overstated because of their key role in recycling nutrients. In addition to decomposers, some fungi are parasites, pathogens, carnivores and mutualists. It is notable that about 80% of all plant diseases are fungal, and 90% of all land plants have mutualistic relationships with fungi.

There are a number of characteristics that unite Kingdom Fungi. These include a cell wall made of chitin, the formation of a spindle inside the nucleus during cell division, the absence of flagellae and centrioles, and a unique **haplontic life cycle** (in most). Fungi also produce **spores** sexually or asexually. Traditionally, four phyla of fungi have been recognized: Chytridiomycota, Zygomycota, Ascomycota and Basidiomycota. These divisions are based largely on differences in their sexual reproduction. A fifth phylum, Deuteromycota, was a garbage-can taxon used for fungi whose sexual reproduction was unknown or non-existent and therefore could not be placed with confidence in one of the other four phyla. This phylum is now largely defunct, but this name is still sometimes used as a term of convenience.

Fungi are predominantly filamentous organisms but a few are unicellular. The fungal body, called a **mycelium**, is composed of thread-like strands of cells called **hyphae**. Hyphae may be **septate**, having one or more perforations in the wall between cells allowing cytoplasm to flow between neighboring cells, or **aseptate/coenocytic**, lacking septae between nuclei (**Fig. 7.1**).

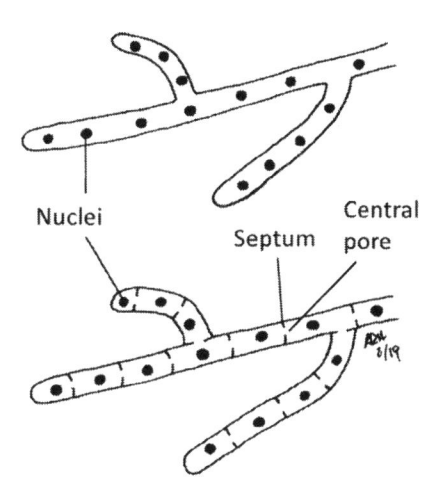

Figure 7.1. Top: Aseptate/coenocytic hypha. **Bottom:** septate hypha. (Image: ARH)

Some common names associated with fungi are not strictly assigned to one phylum or another but instead refer to morphological forms or patterns of growth. For example, a **yeast** is a fungus that is morphologically unicellular and grows asexually by **budding**. A **mold** is a rapidly growing fungus that is reproducing asexually. Any soil fungus that forms a mutualistic relationship with a plant's roots is called a **mycorrhiza**. As noted above, the vast majority of land plants have mycorrhizal associations. A **lichen** is a symbiotic association between a fungus and usually either unicellular green algae or cyanobacteria.

Classification of fungi is based largely on reproductive anatomy and life history. Most fungi reproduce by means of spores produced either asexually or sexually. Sexual reproduction in fungi involves the fusion of hyphae, called **plasmogamy**. This results in hyphae in which each subsequently produced cell contains two nuclei, one from each of the fusing hypha, and is called the **dikaryotic stage** or **dikaryon** (n+n, not diploid). At some later time, the two nuclei inside a dikaryon cell fuse in a process known as **karyogamy** to produce a diploid (2n) **zygote**. Meiosis usually occurs immediately after karyogamy to form haploid spores (see **Fig. 7.2**). Plasmogamy and karyogamy may be widely separated by time or distance. Spores are then dispersed and germinate independently. Each spore can produce a new mycelium. In two groups of fungi, ascomycetes and basidiomycetes, spores generated by karyogamy are produced in specialized structures called **fruiting-bodies**. The fruiting-body is readily recognizable as a mushroom or toadstool. In Ascomycetes the fruiting-body is called an **ascoma** (or ascocarp) and in the basidiomycetes it is called a **basidioma** (or basidiocarp). Zygomycetes and chytrids do not produce complex fruiting-bodies.

By the way, the study of fungi is called mycology. Scientists who study fungi are fond of saying, "Mycology is better than your -cology." (Ok, so, it's a bad joke…)

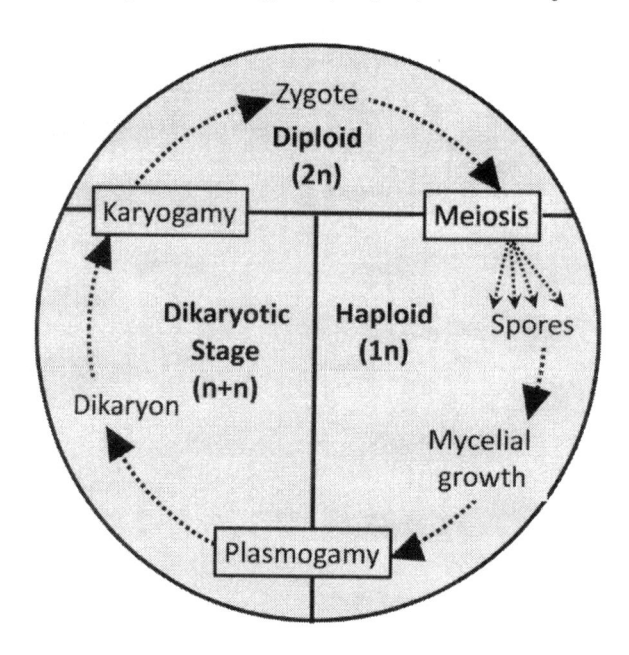

Figure 7.2. Generalized fungal life cycle. Terms in boxes indicate processes by which fungi move from one life stage to the next. (Image: ARH)

Task #1: **Phylum Zygomycota (zygomycetes or zygote fungi; word root = yolk, fungus)**
Rhizopus (bread mold). The hyphae of this fungus are typical of all zygomycetes in that they are aseptate/coenocytic. The material on the prepared slide has been stained blue-green. The slide contains **zygosporangia** that are used in sexual reproduction and **sporangia**, which contain asexually produced spores.

1) Use a compound microscope to observe, describe and DRAW mature zygosporangia of *Rhizopus*. They are the round, dark, opaque structures that have enlarged hyphae (**suspensors**) attached on either side. The zygosporangium will detach from the suspensors and over-winter as a zygospore. Immature zygosporangia are translucent and look like the attached suspensors. Draw and label both a mature and immature zygosporangium and their attached suspensors (hyphae). Refer to the left-hand image on **Figure 7.3** to help you identify what you see.
2) Use a compound microscope to observe, describe and DRAW a sporangium of *Rhizopus*. **Sporangia** that produce spores asexually are attached to the end of a single hypha called a **sporangiophore**. Mature sporangia are round and filled with tiny spores that should appear yellow-brown. The swollen end of the sporangiophore will also be visible. Immature sporangia will lack spores. Draw and label a sporangiophore and its attached sporangium. Refer to the right-hand image on **Fig. 7.3** to help you identify what you see.

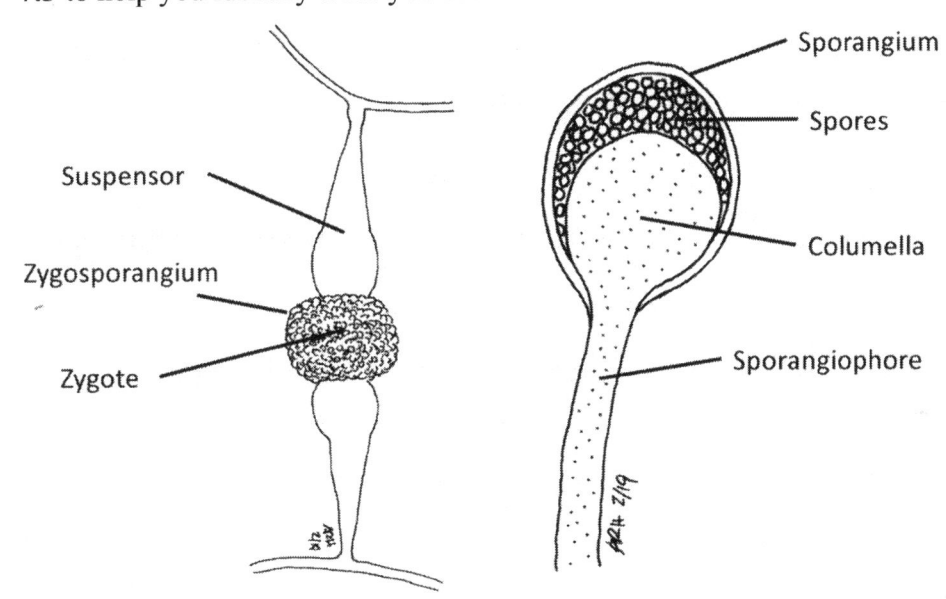

Figure 7.3. Left: Sexual zygosporangium, and **Right**: Asexual sporangium of *Rhizopus*. (Image: ARH)

Task #2: Phylum Ascomycota (ascomycetes or sac fungi; word root = sac, fungus)

Peziza. In ascomycetes, the bowl-shaped **ascoma** is the reproductive structure and is made of a mixture of tightly intertwined **monokaryotic** and **dikaryotic hyphae**. The dikaryotic hyphae form a distinctive layer called the **hymenium** on the inner surface of the ascoma. The hymenium is composed of many **asci** (singular: ascus), which are sac-like structures formed at the tips of the dikaryotic hyphae. Each ascus contains eight **ascospores** which are produced by a zygote that was formed during karyogamy, followed by meiosis and one mitotic division. Some of the ascospores are dark and others are lighter. This is due to genetic differences between them.

1) On lowest magnification, observe, describe and DRAW a portion of the ascoma and label the hymenium on the slide containing a thin section of a bowl-shaped **ascoma** (fruiting body; left-hand image, **Fig. 7.4**).
2) On higher magnification, observe, describe and DRAW the hymenium close-up and label an ascus and its ascospores. Use **Fig. 7.4** to help you identify what you see.

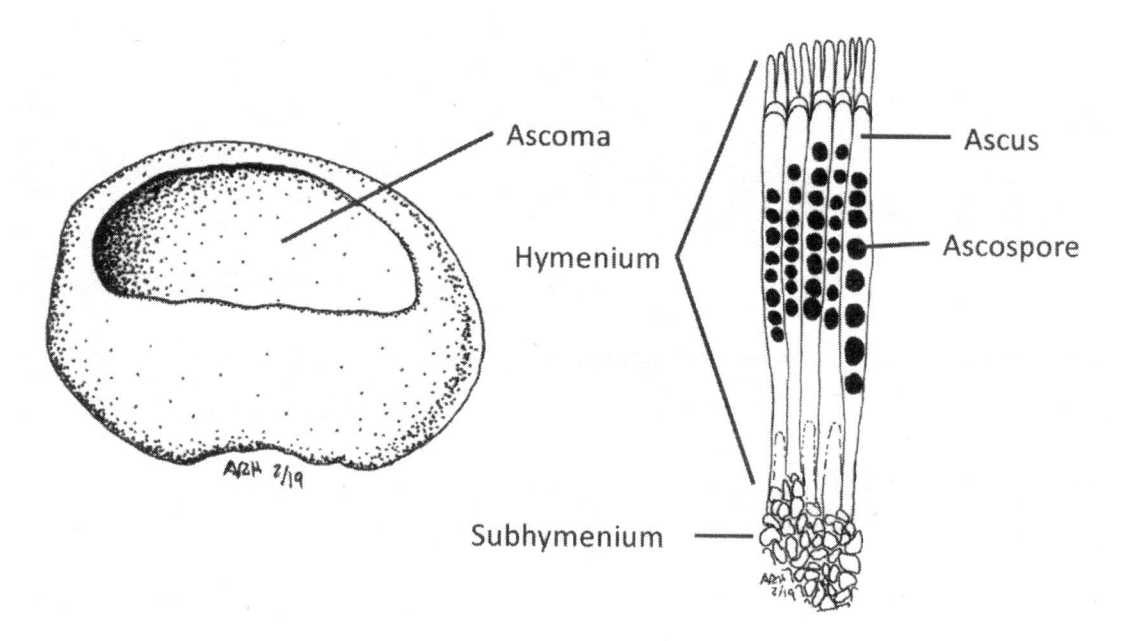

Figure 7.4. Left: Ascoma and **Right:** hymenium of *Peziza*. (Images: ARH)

<u>Task #3</u>: **Phylum Basidiomycota (basidiomycetes or club fungi; word root = pedestal fungus)**

A mushroom or toad stool produced by a member of this group of fungi should be familiar to just about everyone, and is the fruiting body of members of this phylum.

1) Obtain a mushroom (mature **basidioma**) with exposed **lamellae** (gills) and a button mushroom (immature basidioma) that does not yet have the lamellae exposed. These need not be the same species. The underside of the **pileus** of the mature basidioma has exposed, long flaps of tissue called lamellae that extend from the edge of the cap inward toward the **central stipe**. Each gill is coated with a densely packed **hymenium** layer that bears basidia that produce dark brown spores – you will look for these later. Observe, describe and DRAW both of your specimen(s). Be sure to label the following structures: the **pileus**, **stipe**, **veil** (thin layer of tissue covering gills if gills are not exposed), **annulus** (ring around the stipe where the veil was once attached if the gills are exposed). Use **Fig. 7.5** to help you identify what you see.
2) Dissect an immature basidioma (button mushroom) by making a longitudinal cut through the stipe and cap. Observe, describe and DRAW its general morphology, including the stipe (stalk), pileus (cap), veil (membrane on undersurface of pileus, covering the gills), and gills (lamellae).

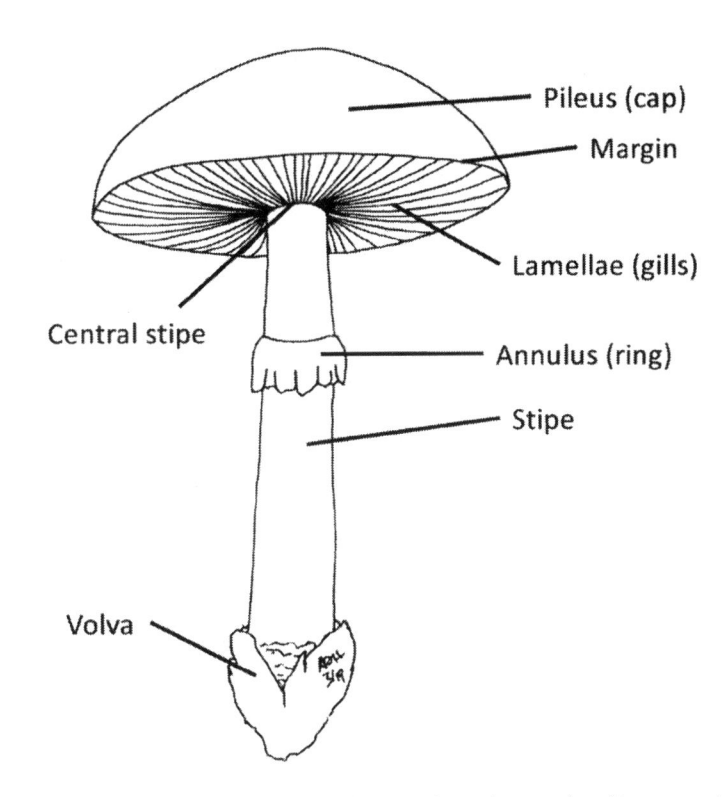

Figure 7.5. General anatomy of a mature basidioma (mushroom). (Image: ARH)

3) Observe the structure of the hymenium. Carefully remove a small piece of one of the lamellae from the mature basidioma and make a wet or dry-mount slide. Observe, describe and DRAW numerous **basidia** that should be visible on the margin of the gill, and the many **basidiospores** attached to them (these will come off easily in a wet mount).
4) Obtain a prepared slide of *Coprinus*. The slide should be a prepared cross-section through the entire pileus (cap) of the **basidioma**. In the Basidiomycota, the basidioma is composed solely of dikaryotic hyphae. The **stipe** (stalk) of the basidioma can be seen in the middle with the **pileus** (cap) forming a ring around it. The inner surface of the pileus has gills that appear as inward projections on the cross-section slide from the margin inward toward the central stipe. Both surfaces of each gill are lined with a thin layer of cells called the **hymenium**. At higher magnification you should see larger blue-stained cells of the hymenium. Some of these extending above the rest and are called **basidia** (singular: basidium). Basidia produce four basidiospores (stained red) each. Spores are connected to the basidium by small pedestals or connections called **sterigma/sterigmata**. Mature spores readily break away from their basidia and are dispersed.
    a. Use the lowest magnification of a compound microscope to observe, describe and DRAW a portion of the **basidioma** and label the **pileus**, **stipe** and **gills**.
    b. Use a higher magnification to observe, describe and DRAW the **hymenium**. Label a **basidium** and its **basidiospores**. Use **Fig. 7.6** to help you identify what you see.

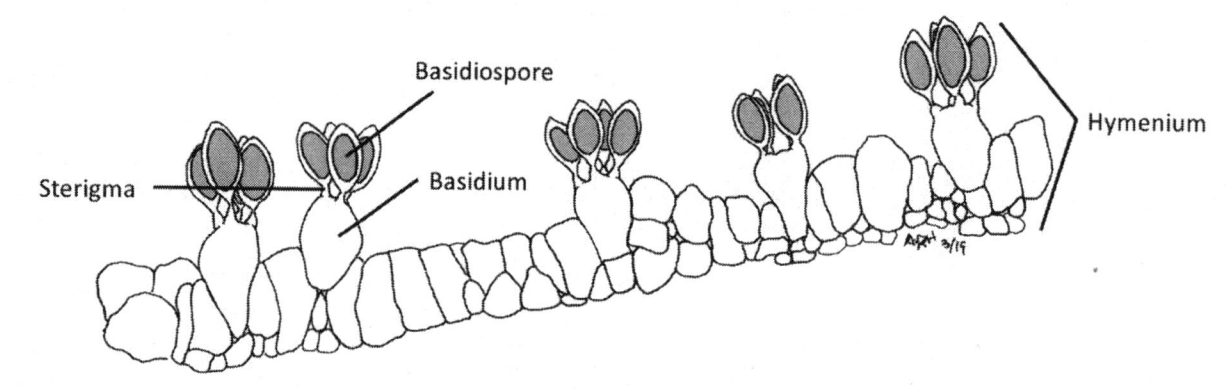

Figure 7.6. *Coprinus,* the hymenium of one side of one gill. (Image: ARH)

Task #4: Phylum Deuteromycota (conidial or imperfect fungi; word roots = second fungus)

Members of this phylum are referred to as the imperfect fungi. They are called imperfect because either their entire life cycle is not yet described or a sexual reproductive phase is missing or has never been described. As mentioned in the introduction to this chapter, this phylum has long been treated as a garbage can phylum and is now largely defunct as a taxonomic unit. It is included here as a matter of convenience. Most deuteromycetes are now known to be asexually reproducing ascomycetes, this connection is based largely on DNA evidence, but others are basidiomycetes or zygomycetes. This group does however contain some extremely important species, a few of which are included here for your examination.

Penicillium. This species is the source of the antibiotic, penicillin. The antibiotic nature of this fungus was reportedly discovered somewhat accidentally in 1928 by Alexander Fleming when fungi contaminated bacterial cultures he was growing. He looked closely at these cultures and observed that there was a bacteria-free halo all the way around the fungus. He concluded, correctly, that the fungus produced a substance that prevented bacteria from growing near it. By the early 1940s, penicillin had been isolated, purified and tested, and began to be used increasingly widely to treat all kinds of infections. It was proclaimed a wonder drug (which it was) and played a significant role during WWII, saving the lives of countless people. Dr. Fleming received the Nobel Prize in 1945 for his discovery of penicillin.

1) *Penicillium.* Obtain a living sample and a prepared slide of *Penicillium.* In this fungus, specialized hyphae called **conidiophores** are branched at the tip, each branch bearing a short chain of **conidia** (i.e., spores) at its tip.
    a. Observe living *Penicillium.* Obtain a sample of *Penicillium roqueforti.* This mold gives bleu cheese its distinctive look, smell and taste. Use a hand lens or dissection microscope to observe, describe and DRAW its conidiophores and conidia, they are grayish green and extremely small.
    b. Observe a prepared slide of *Penicillium.* Use a compound microscope to observe, describe and DRAW conidiophores and conidia of this species. Refer to **Fig. 7.7** to help you identify what you see.

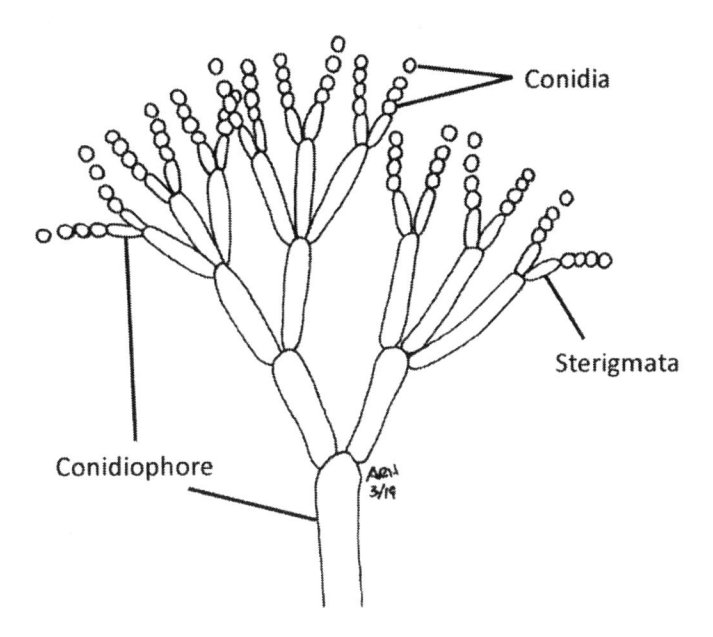

Figure 7.7. Conidiophore and conidia of *Penicillium*. (Image: ARH)

2) *Aspergillus niger*. This species produces the black mold commonly seen on some fruits and vegetables. Though traditionally placed in the Deuteromycota, we now know that it is an ascomycete. This fungus reproduces asexually by means of conidia.
   a. Obtain a fresh sample of *Aspergillus* and examine it using a dissection microscope. It is easiest to see details near the edge of the mycelium. Hyphae in this species are translucent. Use the highest magnification of the dissection microscope to observe, describe and DRAW clusters of conidia and conidiophores. The conidiophore will look like a fine filament tipped by a cluster of conidia that look like tiny spheres. See if you can also find immature conidiophores that have not yet produced conidia.
   b. Obtain a prepared slide of *Aspergillus*. Use a compound microscope to observe, describe and DRAW details of a conidiophore and its conidia. Refer to **Fig. 7.8** to help you identify what you see. Be advised that what you see under the microscope will not look like **Fig. 7.8** because the figure shows a cross-section view of this structure, and you won't see that. It is provided to help you understand the anatomy of these structures.

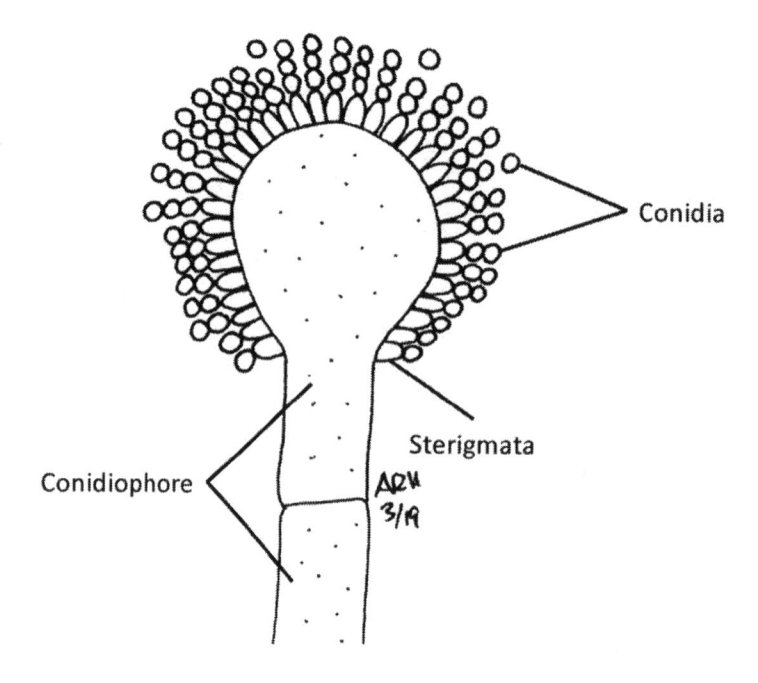

Figure 7.8. Cross-section view of conidiophore and conidia of *Aspergillus*. (Image: ARH)

Task #5: Lichens

Lichens are not a taxonomic group, but are symbiotic associations of fungi and algae. They are included here because you have now been introduced to fungi and algae and you are prepared to understand the nature of this mutualistic association.

In a lichen, a fungus builds a three-dimensional structure made of its hyphae and the algae lives embedded within this structure where it carries out photosynthesis. Some lichens are able to withstand extremely harsh environmental conditions and are important early successional or pioneer species in ecosystem succession. A lichen has one of three basic forms: crustose, foliose and fruticose (see **Fig. 7.9**).

When the fungal component of a lichen is preparing to reproduce it generates structures called **apothecia**. Sometimes they are conspicuous, as shown on the crustose and foliose forms of lichens in **Fig. 7.9**. Apothecia of fruticose forms are often much smaller, requiring careful observation to discover. The structure of a cup-shaped apothecium is shown in **Fig. 7.10**. You are not expected to observe the ultrastructure of one of these structures, but seeing its structure can help you understand how it functions.

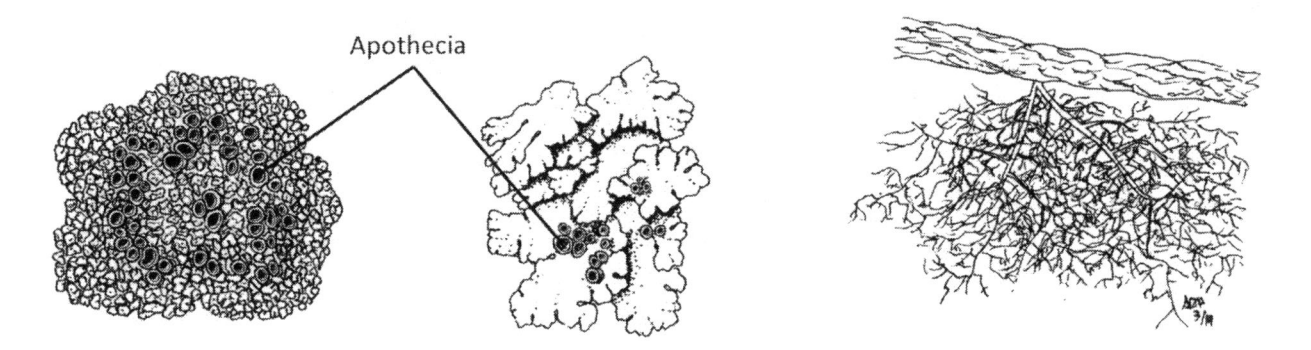

Figure 7.9. Morphological forms of lichens: **Left** - crustose, **Middle** - foliose, **Right** - fruticose. The cup-shaped structures on the crustose and foliose images are apothecia, reproductive structures of the fungal symbiont. Apothecia on fruticose forms are usually extremely small and difficult to see without magnification. (Images: ARH)

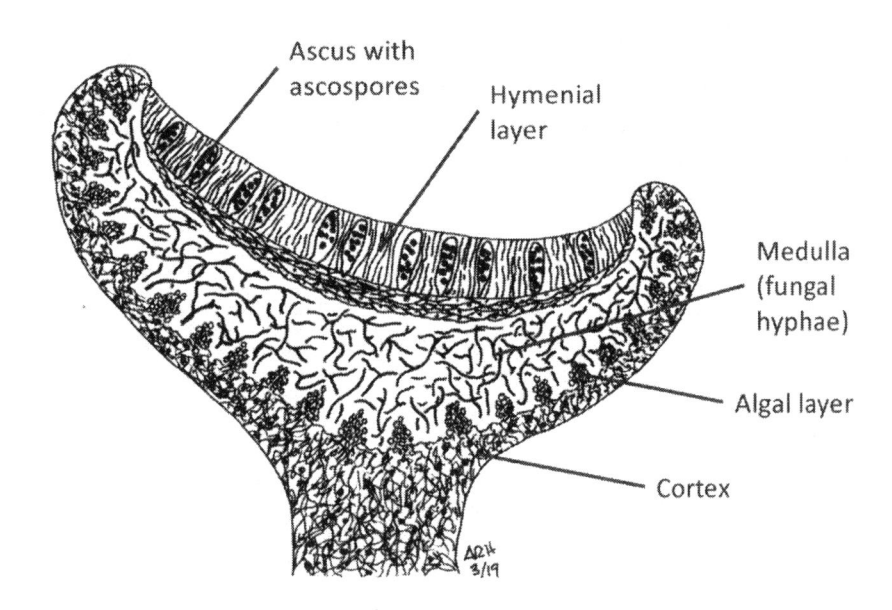

Figure 7.10. Cross-section of a cup-shaped apothecium. (Image: ARH)

1) Living **crustose** lichen. In crustose lichens the **thallus** (lichen body) is tightly adherent to the substrate and cannot be easily removed. These lichens can produce many **apothecia** – the sexual reproductive structures of the fungal symbiont. An apothecium is usually button or cup-shaped, and the interior of the structure bears spores in asci. An apothecium is about the same color as the thallus but appears slightly translucent. See the left image on **Fig. 7.9**.
2) Living **foliose** lichen. In foliose lichens the thallus is not tightly adherent to its substrate and usually contains an upper and a lower cortex. The thallus in usually attached to the substrate by means of spike or hair shaped root-like **rhizines**. Notice that it may have apothecia, and that the thallus has black, marginal setae (hair-like structures) and pores. See the middle image on **Fig. 7.9**.

3) Living **fruticose** lichen. Fruticose lichens often appear shrubby (fruticose means "shrub-like"). They sometimes have a single point of attachment, but not always. Those growing on tree branches are often pendant (hanging). It is not uncommon for people to call this kind of lichen a moss, which is incorrect. In some kinds or fruticose lichens, the thallus may be filamentous or thread-like. Look for apothecia on these lichens. See the left image on **Fig. 7.9**.

4) Choose one of the lichens you examined and DRAW its thallus, indicating its apothecia.

Group Questions

1) Given what you now know about life cycles in general, explain how fungi are able to have such complicated life cycles without having any flagellated or self-motile cells.
2) By now you're probably wondering why there are so many names for spores in Kingdom Fungi. Well, explain why are there so many names for spores in this kingdom.
3) How many different lichens and different kinds of lichens can you find on the sample provided?

Fungi – Glossary

Annulus – flap of tissue around the stipe of a basidioma (mushroom) where the margin of the pileus and the stipe were attached via the veil before separating and exposing the lamellae

Apothecia/apothecium – structure on a lichen in which the fungal symbiont produces spores

Asci/ascus – sac-like structure inside cells of the hymenium of ascomycetes where ascospores are produced

Ascoma – fruiting body of ascomycetes, commonly called cup fungi

Ascospores – sexually produced spores of ascomycetes

Aseptate hypha – see coenocytic hypha

Basidia – cells that produce basidiospores

Basidioma – fruiting body of basidiomycetes, commonly called mushrooms or toad stools

Basidiospore – spores produced sexually by basidiomycetes

Cap – see pileus

Central stipe – portion of the stipe that extends up into the pileus and to which lamellae are attached

Coenocytic hypha – hypha having many nuclei that are not separated from each other by septae

Conidiophore – structure that bears spores in deuteromycetes

Conidium/Conidia – spores produced asexually by deuteromycetes

Crustose – living tightly adhered to and encrusting the surface of a substrate

Dikaryon/dikaryotic stage – hypha made up of cells containing two nuclei that have not yet fused, an n+n condition that is the product of plasmogamy

Dikaryotic hypha – hypha in which all cells have two (n+n) nuclei

Foliose – having a lobed or leaf-like shape

Fruticose – bushy or shrub-like

Fruiting-body – specialized multicellular structure in which spores are produced

Gill – see lamella

Haplontic life cycle – life cycle in which the zygote is the only diploid life stage, and the zygote carries out meiosis immediately after being formed

Hymenium – layer of cells in an ascoma where spores are produced

Hyphae – filaments of cells produced by fungi

Karyogamy – fusion of two nuclei in a dikaryotic cell to produce a diploid cell, i.e., a zygote

Lamella – sheet-like or page-like structures

Lichen – mutualistic association of a fungus and a green alga or cyanobacterium

Mold – a rapidly growing fungus that is reproducing asexually

Monokaryotic hypha – hypha in which all cells have only one nucleus

Mycelium – body of a fungus, is made of many hyphae

Mycorrhiza – fungus that is mutualistic with roots of land plant

Pileus – cap-shaped top of a mushroom (basidioma)

Plasmogamy – fusion of two haploid hyphae to produce a dikaryotic hypha

Rhizines – hair or spike-like structures by which foliose lichens attach to the substrate

Saprobe/saprophytic – organism that secretes digestive enzymes onto its food externally and then takes it up via phagocytosis or pinocytosis

Septate hypha – hypha in which nuclei are separated from each other by cross walls called septae

Sporangiophore – hypha that supports a sporangium in zygomycetes

Sporangium – structure in zygomycetes where spores are produced asexually

Spore – single cell that can germinate and grow independently, does not have to fuse with another cell like gametes do

Sterigma/sterigmata – small pedestal or connection between a cell and a spore it produced

Stipe – central stalk of a basidioma (mushroom)

Subhymenium – cells underlying the hymenium

Suspensors – cells (hyphae) that support the zygosporangium

Thallus – body of a lichen

Veil – thin layer of tissue covering the gills before the margin of the bell and the annulus separate

Volva – remains of the universal veil out of which the entire basidioma (mushroom) grew

Yeast – a unicellular fungus that reproduces by budding

Zygosporangium – cell produced by zygomycetes when complementary hyphae fuse, and in which a dikaryon and then a zygote is formed, followed by meiosis and spore production

Zygote – diploid cell produced when two haploid cells fuse or when nuclei of a dikaryotic cell fuse

Chapter 8: Animalia - Basal Animals

Kingdom Animalia has well over one million described species of animals, and about 30 different phyla. Each phylum is characterized by its own unique body plan, each with its own set of ecological advantages and challenges. The vast majority of described animal species belong to the following phyla: Porifera, Cnidaria, Platyhelminthes, Mollusca, Annelida, Nematoda, Arthropoda, Echinodermata and Chordata.

Efforts to classify animals go back about 2,500 years to Aristotle, in ancient Greece. You would probably think that by now we should have answered all the questions there are to ask about animal classification. We have answered many such questions, but answers to a few major questions still elude us. One such question, that has been the subject of intense debate for the past decade is, which is the basal animal phylum? That is, which group of animals first branched off of the animal evolutionary tree? There are currently four candidates for this distinction: Phylum Ctenophora (comb jellies), Phylum Porifera (sponges), Phylum Cnidaria (jellyfish and relations), and Phylum Placozoa (no common name) (see **Table 8.1**).

Table 8.1. Basal animal phyla.

Phylum name	Common name or representative	Habitat and ecology	Image
Ctenophora	Comb Jellies	Marine planktonic and benthic predators	
Porifera	Sponges	Marine and freshwater benthic suspension feeders	
Cnidaria	Jellyfish, sea anemones, coral and relations	Marine and freshwater benthic and planktonic predators	
Placozoa	Placozoans	Marine benthic opportunistic heterotrophs	

For many years, sponges were accepted as the basal animal phylum. They are anatomically simple and some of their cells bear a striking similarity to a group of protozoans called choanoflagellates that are currently accepted as the sister taxon to Kingdom Animalia. Recent discoveries have cast doubt on this conclusion, and because of growing, compelling evidence, most zoologists now accept the conclusion that ctenophores, not sponges are the basal animals.

In this exercise you will be introduced to three animal phyla that are contenders for the position of basal animal phylum: Phylum Ctenophora, Phylum Porifera, and Phylum Cnidaria. A fourth phylum, Placozoa, is not presented.

Placozoans are tiny animals up to about 1mm across and made of hundreds to thousands of cells, and they all belong to a single genus. At first glance placozoans look and move like large amoebae. Phylogenetic analysis indicates that placozoans are not the basal animal phylum. Placozoans have dorsal and ventral ciliated epithelial layers with a central space housing stellate cells reminiscent of mesenchyme cells of other animals.

Phylum Ctenophora (ctene = comb, fera = bearing)

This is one of the so-called minor phyla, but has been the focus of much research during the past decade, because there is a growing body of evidence that ctenophores, commonly called comb jellies, are, surprisingly, the basal animal phylum. There are about 150 described species of ctenophores. Ctenophores are common members of marine planktonic and benthic communities. Interestingly, these are the largest animals that move solely via ciliary action. All ctenophores are carnivores. They use sticky **colloblast cells**, lobe-like appendages, or large mouths to capture and ingest prey. These animals are diploblastic, possibly **triploblastic**. They have **biradial symmetry** and a branching **gastrovascular cavity**. One of their unique features is their **ctene rows**. Each ctene is made up of 100s of cilia fused together to form a sort of a paddle. Ctenophores beat their ctenes in a coordinate pattern to produce locomotion.

The placement of ctenophores within the animal tree of life has long been a mystery. In the past, some scientists proposed that ctenophores and cnidarians form a clade. Others asserted that ctenophores are a stand-alone phylum. Recent detailed analysis of the anatomy, development and molecular biology of ctenophores has led to a well-supported conclusion that ctenophores are a stand-alone phylum and that they are the **basal taxon** of animals. Why? Ctenophores apparently lack the same *Hox* genes – developmental regulatory genes – possessed by all other animals. They also have neurogenic, immune system and ribosomal protein genes that are different than those of other animals, suggesting that ctenophores are the basal animal phylum.

<u>**Task #1:**</u> **Ctenophore anatomy**

1) Submerge and examine a preserved ctenophore. It will most likely be *Pleurobrachia*, also called the sea gooseberry or sea walnut. Observe, describe and DRAW what you see. Use **Fig. 8.1** and **Fig. 8.2** to help you identify structures on your specimen. Note: The tentacles were probably retracted into their tentacle sheaths when the animal was preserved.

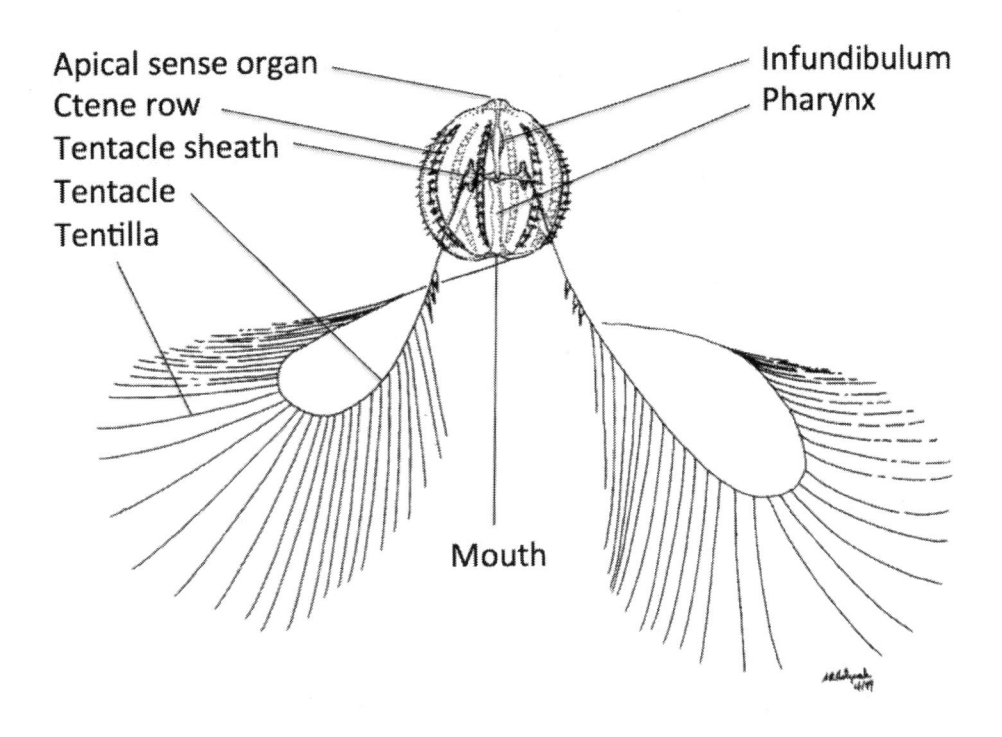

Figure 8.6. A sea gooseberry, the ctenophore *Pleurobrachia,* with tentacles and tentilla extended. (Image: ARH)

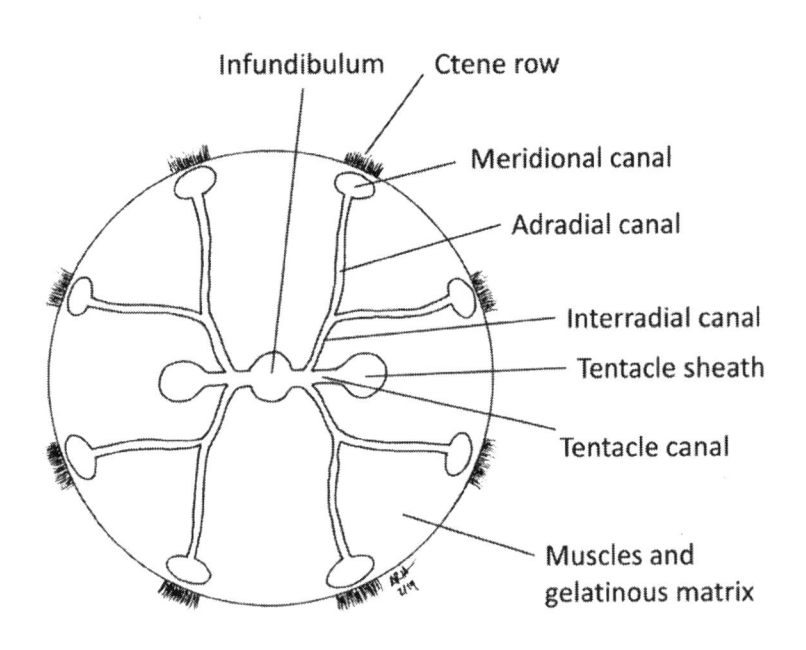

Figure 8.7. Internal canal system of a ctenophore like *Pleurobrachia*. (Image: ARH)

Phylum Porifera (word roots: pori = pore, fera = bearing)

There are over 5,000 species of sponges and all but 150 species are marine. All sponges are **suspension feeders**. Some sponge cells are specialized for capturing tiny particles from the water but every sponge cell that is in contact with the water can phagocytize anything that comes in contact with it.

Sponges differ from other animals in that the division of labor takes place at the level of specialized cells rather than among specialized tissues or organs. Sponges lack true tissues and don't have any organs. In most animals the outermost layer of cells of the body forms an epithelial layer that plays an essential role in isolating the external environment from the internal environment of the animal and controlling what can pass across it. This isolation is accomplished by a true epithelial tissue with a variety of cell junction molecules called cadherins that connect neighboring cells to each other. Some kinds of cadherins form cell junctions called gap junctions that allow neighboring cells to communicate directly with each other. Sponges have a few kinds of cadherins but they lack others and do not form cell junctions, so the outer layer of sponge cells does not restrict movement of materials across it like the epithelia of other animals.

There are three basic body plans observed among cellular sponges: **asconiod**, **syconoid** and **leuconoid** plans. You need to know that these body plans are not good taxonomic characters and may be found across various taxa, but they are useful in understanding sponge anatomy. The asconoid and syconoid plans are sac or tube shaped. In the asconoid plan the body wall is a straight tube and in the syconoid plan the body wall has many out-pocketings. The leuconoid plan however is much more complex and has a series of branching incurrent canals that lead to small feeding chambers and water leaves these via excurrent canals that eventually exit the body. See **Fig. 8.3** to see diagrams depicting these body plans.

Task #2: Body plan of an asconoid sponge

1) Obtain a small sample of a preserved specimen of an asconoid sponge such as *Leucosolenia*. Immerse your specimen in a glass bowl and observe it using a hand-held lens or dissection microscope. DESCRIBE what you see. A drawing may be helpful but is not mandatory. Use **Fig. 8.4** To help you identify what you see.

Task #3: Spongin and spicules

1) Use a compound microscope to observe either a prepared slide or a wet-mount slide of spicules. Observe, describe and DRAW what you see.
2) Make a wet-mount slide of the tiniest piece of spongin possible. Observe, describe and DRAW what you see.

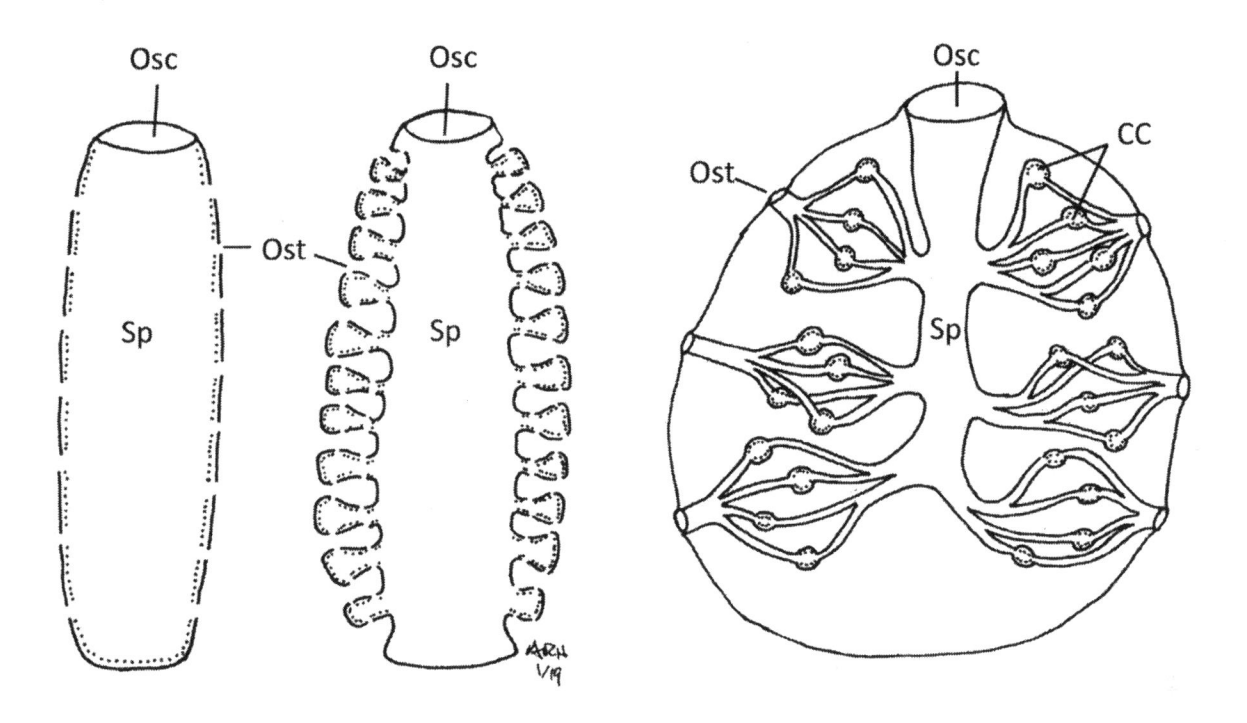

Figure 8.3. Three sponge body plans. **Left:** asconoid. **Middle:** syconoid. **Right:** leuconoid. A series of incurrent canals lead to choanocyte chambers in leuconoid sponges, and a series of excurrent canals carries water to the spongocoel. Dots lining inner surfaces and other locations within sponges indicate locations of choanocytes. Labels: CC (**choanocyte chamber**), Osc (**osculum**), Ost (**ostium**), Sp (**spongocoel**).

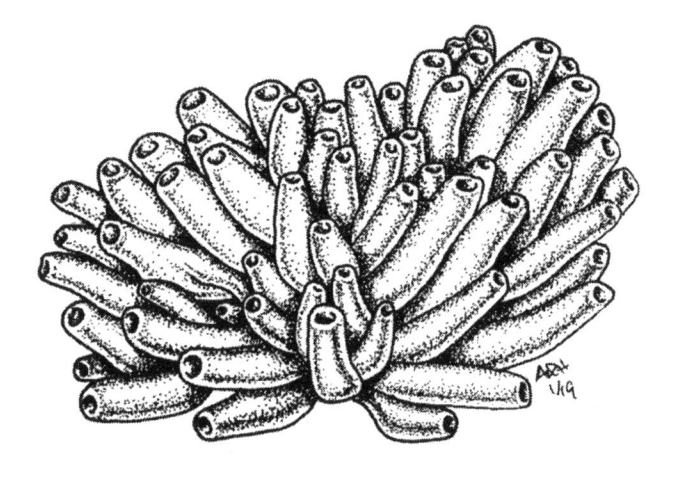

Figure 8.4. An asconoid sponge. The opening at the tip of each tube is an osculum. (Image: ARH)

Task #4: Sponge body structures and cell types

1) Use resources available to you, such as your textbook, to IDENTIFY and LEARN the

functions of the following sponge anatomy, as well as their locations in a sponge body.
   a. **Atrium/Spongocoel**
   b. **Mesohyle**
   c. **Osculum/Oscula (plural)**
   d. **Ostium/Ostia (plural)**
   e. **Spicule**
   f. **Spongin**
2) Use resources available to you to IDENTIFY and LEARN the functions of the following kinds of sponge cells as well as their locations in the body wall of an asconoid sponge.
   a. **Archaeocyte/Archeocyte**
   b. **Choanocyte**
   c. **Myocyte**
   d. **Pinacocyte**
   e. **Porocyte**
   f. **Sclerocyte**
   g. **Spongocyte**
3) PRODUCE a drawing of a small section of the body wall of a sponge and indicate the location of each of the cells and anatomical structures listed above.

Phylum Cnidaria (cnida = a nettle)

There are over 13,000 species of cnidarians including jellyfish, corals, sea anemones and their relations. The cnidarian body plan is anatomically more complex than sponges, but less complex than most other kinds of animals. Cnidarians are **diploblastic**, having only two true tissue layers, the **epidermis** and the **gastrodermis**, with an acellular gelatinous **mesoglea** layer sandwiched between them. Their bodies generally exhibit **radial** or **bi-radial symmetry** and they have a nervous system and muscles.

All cnidarians are carnivores. They capture food using cells called **cnidocytes** that house stinging organelles called **nematocysts**. Food is digested in sac-like gut called the **gastrovascular cavity** that has only one opening that serves as both mouth and anus.

The basic cnidarian life cycle has a sexually reproductive swimming **medusa** stage and a cloning sessile **polyp** stage.

Task #5: Anatomy of the cnidarian body plan

1) Observe a prepared cross-section slide of *Hydra* to see the diploblastic nature of the body wall. Describe and DRAW what you see. Use **Fig. 8.5** To help you identify what you saw.
2) Observe either a prepared slide or preserved specimen of an *Obelia* or similar medusa stage. DRAW what you see, and use **Fig. 8.6** to help you identify what you see.
3) Observe a prepared slide of the clonal polyp stage (colony) of *Obelia* and describe and DRAW what you see. Use **Fig. 8.7** to help you label your drawing.

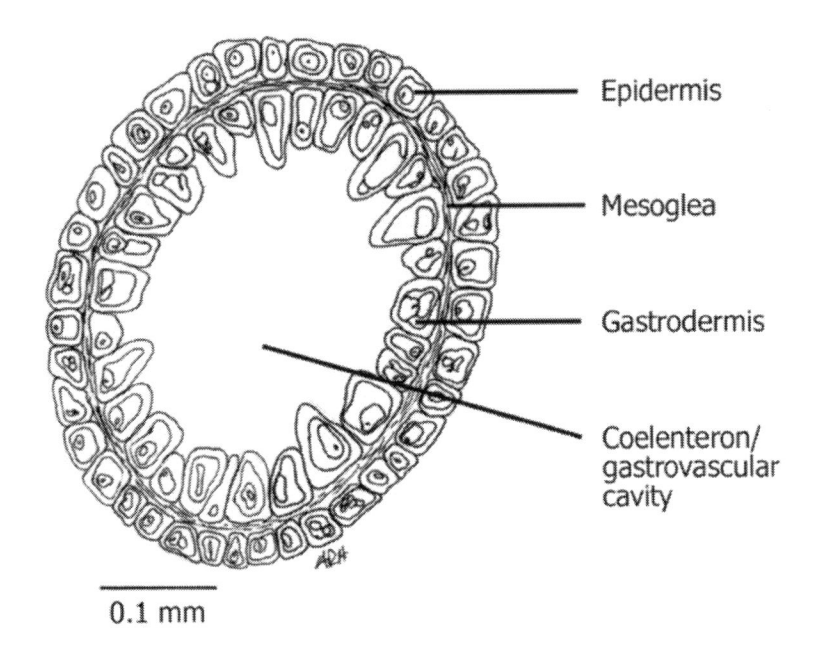

Figure 8.5. Cross-section of the hydrozoan *Hydra*, showing the diploblastic body plan of cnidarians. (Image: ARH)

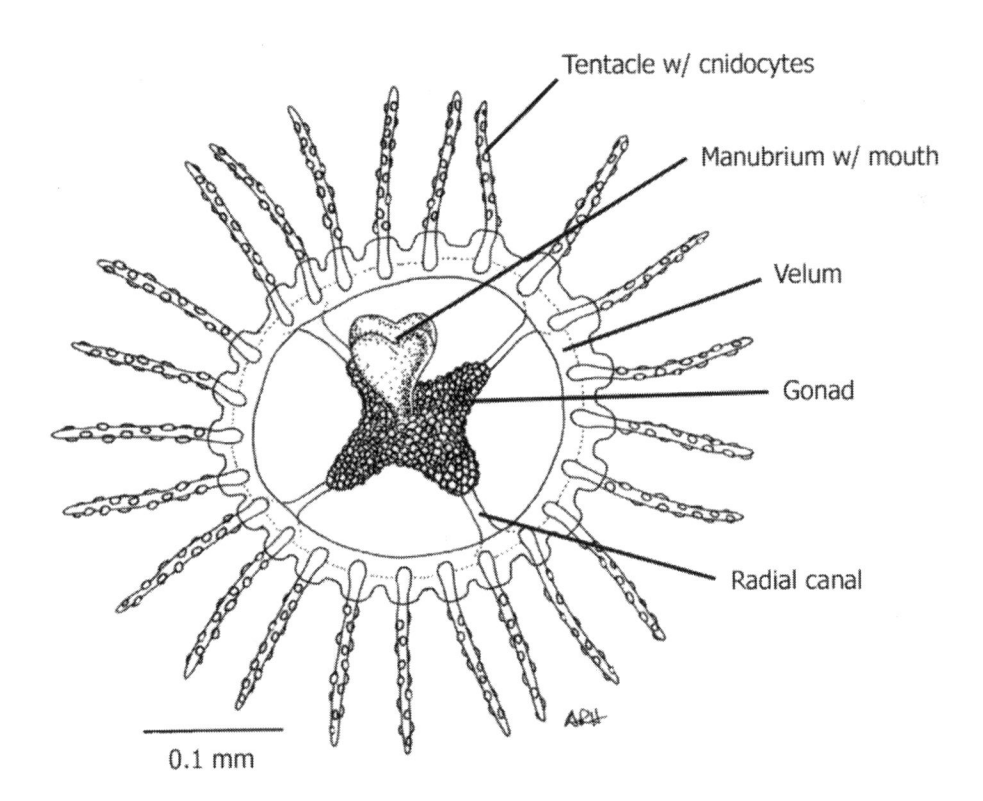

Figure 8.6. *Obelia* medusa stage. (Image: ARH)

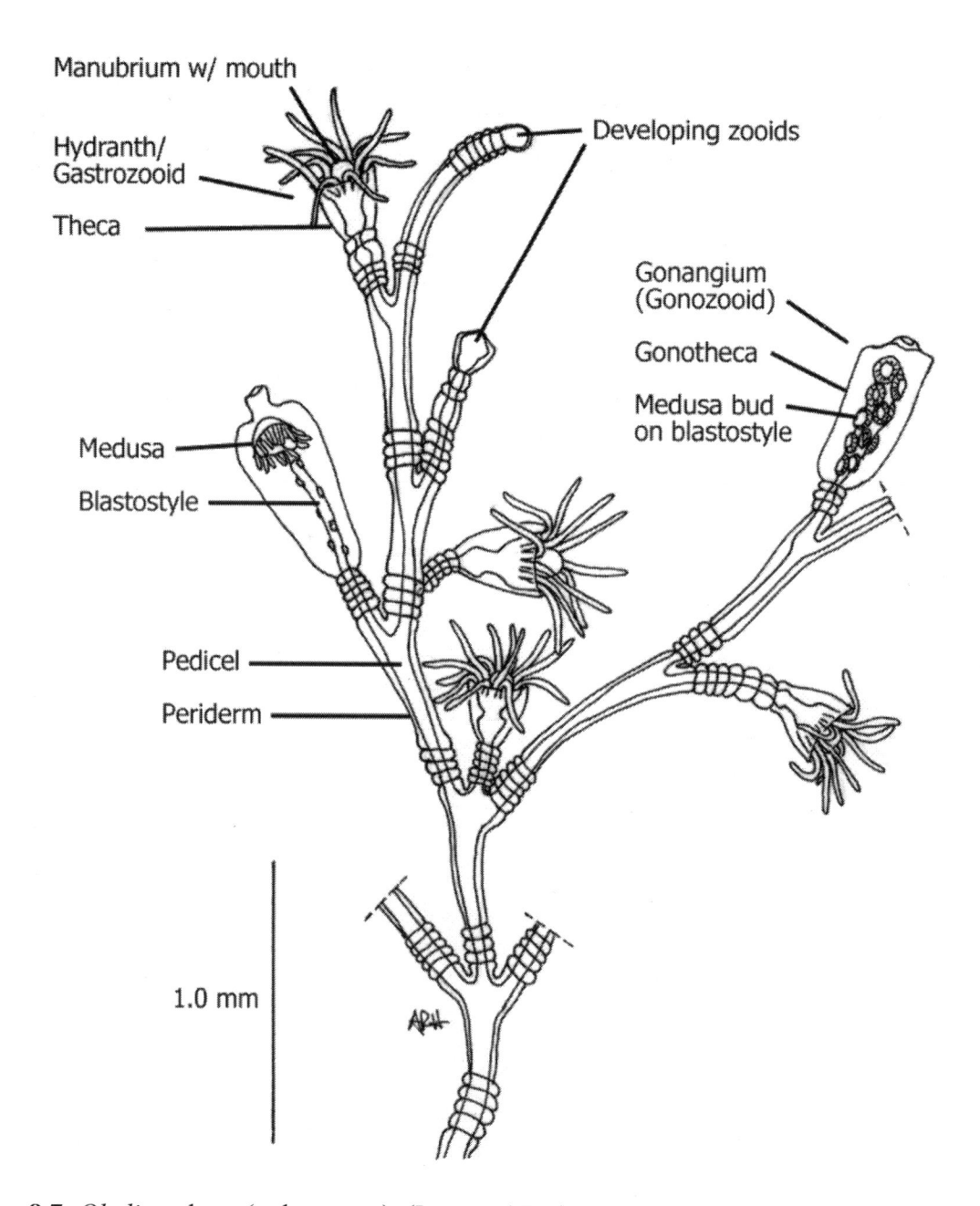

Figure 8.7. *Obelia* colony (polyp stage). (Image: ARH)

Group Questions

1) What characteristics allow sponges to be successful today, even when all other animals are so much more anatomically complex and behaviorally active?
2) Describe one major advantage and one major disadvantage of being radially symmetrical, as displayed by cnidarians.
3) Develop an explanation about why ctenophores were not, at least at first glance, obvious candidates to be the basal animal phylum.

Phylum Ctenophora – Glossary

Adradial canal – canal that carries water, food and oxygen to the meridional canal

Apical sense organ – structure containing a statocyst that regulates beating of the ctene rows, also includes ciliated sensory fields presumably for monitoring water quality

Colloblast cell – sticky cell used to capture prey from the water

Ctene – A plate of 100s of cilia fused together and beating together so they function more like an oar than a whip

Ctene row – a row of many ctenes that beat synchronously to provide propulsion

Infundibulum – a hollow structure that connects the aboral sense organ to other parts of the body

Interradial canal – canal that carries water, food and oxygen to the adradial canal

Meridional canal – canal that carries water, food and oxygen to the cells of the ctene rows

Pharynx – muscular tube that moves food from the mouth to the branched gastrovascular cavity

Tentacle – primary filamentous extension of the body, bears many smaller tentilla

Tentacle canal – canal that carries water, food and oxygen from the infundibulum to the rest of the canal system

Tentacle sheath – cavity into which the tentacles can be withdrawn entirely, these sheaths are not connected to the digestive tract

Tentilla – extremely fine tentacle-like structures that extend off of a tentacle, these bear sticky colloblasts and are the main food capturing structures in tentacle-bearing ctenophores

Porifera - Glossary

Archaeocyte – a pluripotent or multipotent cell found in the mesohyle or a gemmule

Atrium (spongocoel) – central cavity of the sponge body that carries water to the osculum

Choanocyte – a cell bearing a single flagellum surrounded by a collar of microvilli, these are similar in structure and function to choanoflagellates

Choanocyte chamber – a space lined by flagellated choanocytes

Incurrent canal – canals carrying water from ostia to choanocyte chambers

Mesohyle – a gelatinous matrix located between the outer and inner body surfaces of sponges

Oscula (plural) / Osculum (singular) – excurrent opening(s) of a sponge

Ostia (plural) / Ostium (singular) – incurrent opening(s) of a sponge

Pinacocyte – a flat cell that covers external and internal surfaces of sponges that are not lined by choanocytes

Sclerocyte – cells that secrete spicules

Spicules – secreted $CaCO_3$ or SiO_2 structures that provide skeletal support and physical defense in sponges and protection to gemmules

Spongin – stiff yet flexible supporting material made of collagen

Spongocoel (atrium) – excurrent cavity where water collects prior to leaving a sponge body

Spongocyte – cells that secrete spongin

Suspension feeder – feeding by capturing small particles or organisms suspended in the water

Phylum Cnidaria – Glossary

Biradial symmetry – body plan that can be divided into equal halves in only two planes, because the mouth and pharynx are oblong structures within a radially symmetrical body

Blastostyle – rod-like structure in the center of a hydrozoan gonozooid where medusa buds are produced

Cnidocyte – cell that produces a stinging nematocyst

Coelenteron/gastrovascular cavity – fluid-filled space within the body where digestion

Diploblastic – body plan that produces only ectoderm and endoderm embryonic tissues

Epidermis – tissue derived from embryonic ectoderm that covers the outer surface of the body

Gastrodermis – tissue derived from the embryonic endoderm that lines the gastrovascular cavity

Gonangium (gonozooid) – hydrozoan zooids that produce hydromedusae by budding

Gonotheca – chitinous covering that protects gonozooids

Gonozooid – zooid that carries out reproduction in the Portuguese man o' war, also a structure that produces medusa buds in *Obelia* by cloning

Hydranth/gastrozooid – feeding zooids of hydrozoans

Manubrium (with mouth) – tube of tissue bearing the mouth in some cnidarians

Medusa – free-swimming life stage that carries out sexual reproduction

Medusa bud – small button-shaped mass of tissue that pinches off and swims away as an immature hydromedusa in hydrozoans

Mesoglea – gelatinous matrix of material derived from the ectoderm that forms a layer between the epidermis and gastrodermis

Nematocyst – stinging organelle that contains a coiled hollow tube that everts when it fires, penetrating its target and injecting toxins into it

Pedicel/coenosarc – tissue that connects members of a hydrozoan colony to each other

Periderm – transparent chitinous covering that surrounds and protects the pedicel

Polyp – sessile, usually clonal life stage of a cnidarian

Radial symmetry – body plan with a central axis that can be divided into many different mirror image halves

Tentacle – cnidocyte-bearing structures used for feeding, the size of tentacle reflects prey size

Theca – transparent chitinous covering of zooids in hydrozoans

Velum – inward-facing shelf of tissue constricting the subumbrellar opening in hydromedusae, allows the animal to produce an increased velocity of water flow each time the bell contracts

Chapter 9: Animalia - Lophotrochozoa

All animals other than the basal taxa are **triplobastic**, have **bilateral symmetry** and belong to Clade Bilateria. Clade Bilateria is subdivided into two clades: Clade Protostomia and Clade Deuterostomia. Historically, animals were assigned to these two clades based on their pattern of developmental (See **Table 9.1**). It turns out that these developmental traits are not as important for classification as we once thought. These characters are included here though because they are still used widely to demonstrate differences between many protostomes and deuterostomes.

Table 9.1. Developmental differences between protostome and deuterostome animals.

Characteristic	Protostome	Deuterostome
Cleavage pattern	Spiral	Radial
Developmental fate of blastomeres produced during cleavage	Determinate	Indeterminate
Fate of blastopore	Mouth	Anus
Source of mesoderm	4d cell (in most)	Archenteron cells
Pattern of coelom formation	Schizocoely	Enterocoely

Clade Protostomia is subdivided into Clade Lophotrochozoa, the focus of this chapter, and Clade Ecdysozoa, the topic of lab10.

Clade Lophotrochozoa (lopho = crest; trocho = wheel; zoa = animal) is a large group of animal phyla that have bilateral symmetry and use cilia for locomotion, feeding or both. The clade name comes from the fact that some phyla in the clade bear a structure called a **lophophore**, and some have a **trochophore** larval stage. Both of these structures have bands of cilia used for swimming or feeding. There are 14 phyla in Clade Lophotrochozoa (see **Table 9.1**), but only three of them, Phylum Platyhelminthes, Phylum Annelida and Phylum Mollusca are included in this exercise. Most of the other phyla contain only a few species or are microscopic.

Table 9.1. Phyla in Clade Lophotrochozoa. Major phyla are indicated by an asterisk.

Phylum name	Common name or representative	Habitat and ecology	Image
Platyhelminthes*	Flatworms including planarians, flukes, tapeworms	Endoparasites or terrestrial and aquatic predators	

Rhombozoa	Rhombozoans	Microscopic marine endoparasites of cephalopods	
Gastrotricha	Gastrotrichs	Microscopic marine or freshwater herbivores and detritovores	
Entoprocta	Entoprocts	Tiny marine suspension feeders	
Cycliophora	Cycliophorans	Microscopic marine suspension feeders, found only on setae of lobster mouth parts	
Nemertea	Ribbon worms	Voracious marine predators from the intertidal to deep sea	
Annelida*	Segmented worms, spoon worms, peanut worms	Terrestrial, freshwater and marine, every possible lifestyle	
Mollusca*	Snails, clams, chitons, octopus and relations	Terrestrial, freshwater and marine, every possible lifestyle	
Gnathostomulida	Jaw worms	Microscopic opportunistic feeders in marine sediments	

Micrognathozoa	Micrognathozoans	Microscopic opportunistic feeders in freshwater springs in Greenland	
Rotifera	Rotifers	Marine, freshwater and semi-terrestrial microscopic opportunistic feeders	
Phoronida	Horseshoe worms	Suspension feeders in marine soft sediments	
Bryozoa	Moss animals	Marine and freshwater suspension feeders	
Brachiopoda	Lamp shells	Marine suspension feeders	

Phylum Platyhelminthes (platy = flat, helminth = worm)

There are about 27,000 described species of flatworms. This phylum includes free-living turbellarians and **endoparasitic** flukes and tapeworms. Turbellarians are largely opportunistic chemosensory hunters and scavengers, but medically important parasitic flukes and tapeworms occupy the vast majority of scientific attention. About three quarters of all described flatworms are parasites.

The classification scheme of platyhelminths is under revision, so it will not be addressed here, but these traditional flatworm class names will be used instead for convenience:

- Class Turbellaria – free-living flatworms
- Class Trematoda also called Class Digenea – flukes
- Class Monogenea – monogeneans
- Class Cestoda – tapeworms.

Class Turbellaria (turb = to stir up or whirlpools)

This group contains worms that live in marine, freshwater and terrestrial habitats. These flatworms all have three or more main branches of the **gastrovascular cavity**. Freshwater and terrestrial species typically have two branches of the gut extending posteriorly and one anteriorly from the middle of the body where the mouth is located. Marine species have many extensions radiating away from the main lumen of the gut.

You might be surprised to learn that there are large terrestrial flatworms and they are fairly common in regions where it is moist and warm. **Fig. 9.1** shows a terrestrial arrowhead flatworm seen crawling on an exterior window in western Arkansas, USA. These worms are harmless to humans; earthworms are their main prey.

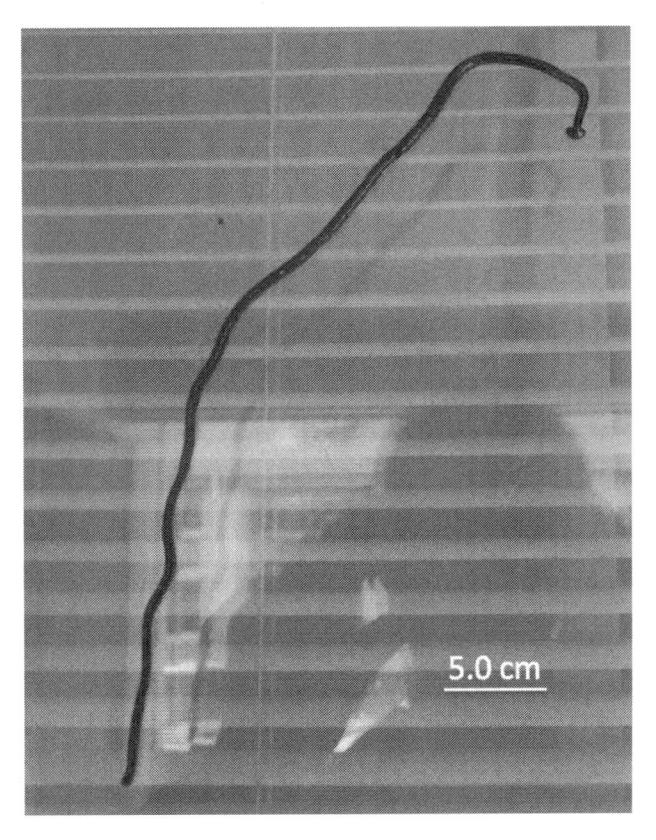

Figure 9.1. Arrowhead flatworm, Arkansas, USA. (Image: ARH modified photo Courtesy of Ryan Durrant)

Class Trematoda (trema = a hole) or **Class Digenea** (di = two, gen = birth)

Trematode flukes have complex parasitic life histories, i.e., they have life histories that include at least two hosts, sometimes more than that. The intermediate hosts in these life cycles are typically molluscs and the final hosts, where adult parasites live, are vertebrates. Interestingly, the infective stages of flukes in intermediate hosts can change the behavior of their hosts in order to increase the likelihood of a parasite finding its way into a suitable host and completing its life cycle.

Class Monogenea (mono = one; gen = birth)

Monogenean flukes are **ectoparasites** and have a simple life history. That is, they have only one host during their lives. Adults release fertilized eggs or larvae that swim in the water and then attach to a host, usually a fish or amphibian. The adult pierces thin epithelial tissue, like the gills or skin on a fin, and they ingest blood and fluids from the host.

Class Cestoda (cestus = a belt or a girdle)

Tapeworms have highly modified bodies, so much so that adult tapeworms do not have a mouth or gut. Instead, their outer body covering is adapted to absorb predigested materials from the intestines of their hosts. These worms have an organ called a **scolex** that bears suckers, hooks or both and is used to attach the worm to a host's intestinal wall. The rest of the body is comprised of a few to many repeated structures called **proglottids** that results in a long strap-like body. Each proglottid has both female and male reproductive organs, and can produce huge numbers of embryonated eggs. Tapeworms, like trematode flukes, have complex life histories though their intermediate and definitive hosts are usually both vertebrates.

<u>Task #1:</u> **Body plans of flatworms**

1) Observe a prepared slide of a turbellarian flatworm that has been stained to highlight the gut. DRAW this worm and WRITE your thoughts about why you think the gut is so large and so extensively branched. Use **Fig. 9.2** to help you understand what you see.
2) Observe a prepared slide of a planarian that clearly shows the eyes. DRAW the head and eyes. Do a little research and explain why they seem to look cross-eyed.
3) Observe a prepared slide of the adult of the trematode fluke *Clonorchis sinensis,* the Chinese liver fluke. Use **Fig. 9.3** to help you identify what you see. WRITE your thoughts about why you think so much of the space inside the body is devoted to reproductive organs.
4) Observe a prepared slide of the tapeworm *Taenia* and look at the scolex. DRAW what you see. WRITE your thoughts about how you think this structure works.
5) Observe a prepared slide of the tapeworm *Taenia* and look at a mature proglottid. Use **Fig. 9.4** to help you identify what you see. WRITE your thoughts about why each proglottid is packed with reproductive organs.

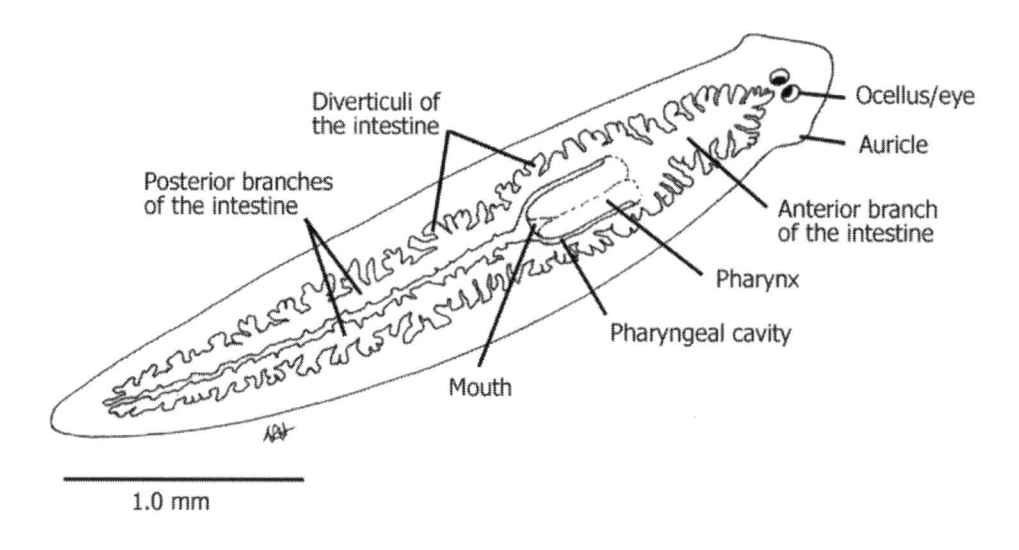

Figure 9.2 The digestive tract and eyes of a planarian flatworm. (Image: ARH)

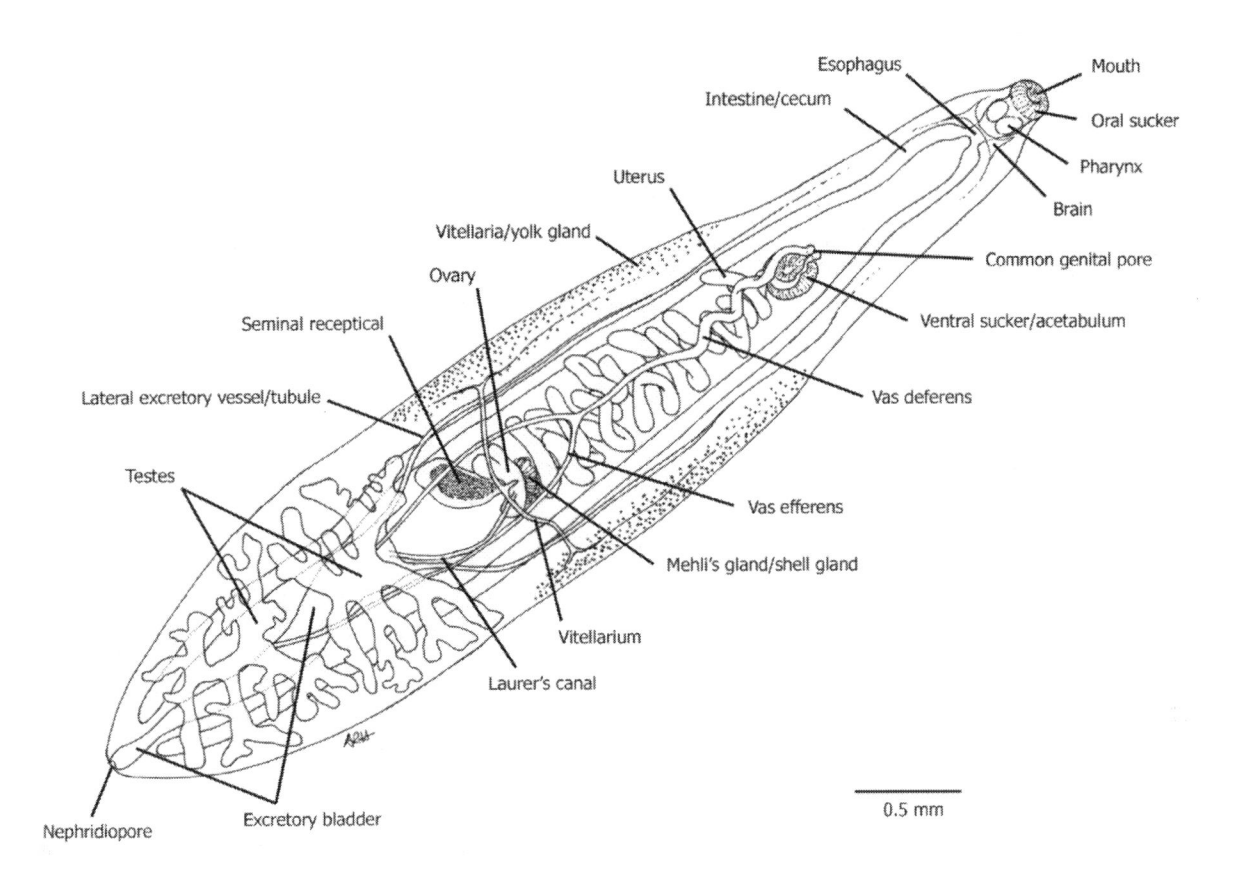

Figure 9.3. Chinese liver fluke, *Clonorchis sinensis*, adult stage. Embryonated eggs in the uterus are not shown to allow increases anatomical clarity. (Image: ARH)

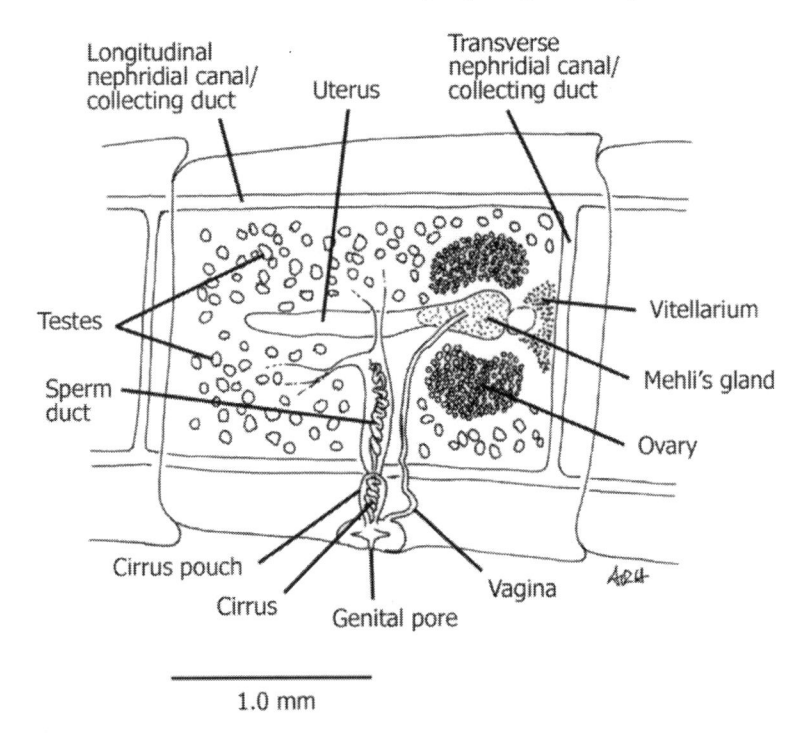

Figure 9.4. Mature proglottid of the tapeworm *Taenia*. (Image: ARH)

Phylum Annelida (anulus = little ring)

Phylum Annelida includes about 20,000 species, including segmented worms and their relatives. The phylum name refers to the ring-like appearance of segments in the traditional annelid body plan. The taxonomy of this phylum has experienced some interesting changes. The annelids used to include only earthworms, marine polychaete worms and leeches. Phylogenetic analysis research has shown however that some animals that formerly had phylum-level status are actually members of Phylum Annelida: sipunculids (peanut worms), echiurans (spoon worms), pogonophorans (beard worms), and vestimentiferans (giant hydrothermal vent worms).

The ancestral traits of annelids include a segmented body, **metamerism**, and bristles called **chaetae**. They also have a closed-circulatory system, fluid-filled coelomic spaces, and **trochophore larvae** (when larvae are present in the life cycle). For convenience, we will consider only the following groups.

- **Class Oligochaeta:** terrestrial and freshwater annelid worms that have few short chaetae, e.g., earthworms.
- **Class Polychaeta**: mainly marine annelids that may be either tube-dwelling suspension feeders, deposit feeders, active predators or scavengers. These have many long chaetae.
- **Class Hirudinea**: the leeches, a group of ectoparasites and predators that feed on the blood of their prey.

Task #2: Earthworm behavior

1) Obtain a live earthworm and observe its behavior. Rinse your specimen with water and place it on a moist paper towel so you can observe its movement. WRITE your observations about how earthworms move.
2) WRITE your thoughts about how this method of movement helps earthworms to be effective burrowers.
3) Discover the location of chaetae on the earthworm by gently running your fingers along the length of the worm from anterior to posterior and then from posterior to anterior. Repeat this for dorsal and ventral surfaces of the worm. WRITE down your observations. DRAW a simple cross-section sketch of the body and show where the chaetae are located.
4) Rinse and place a worm on a dry paper towel and listen carefully as the worm moves. Listen for the scratching sound of chaetae on the paper towel. Do not leave the worm on a dry paper towel too long because it can dry out quickly.

Task #3: Earthworm body plan

1) Obtain a preserved earthworm and use pins to attach the worm, ventral side down, to the wax bottom of a dissection pan (you should have already discovered that chaetae are on the ventral surface). Add enough water to the pan to cover your specimen. Cut through the dorsal body wall and gently pull the body wall open and pin it to the wax tray bottom.

Continue cutting and pinning until you have exposed the anterior portion of the internal anatomy of your worm, as shown in **Fig. 9.5**.

2) DRAW your worm and use **Figure 9.5** to help you identify what you see.
3) WRITE your thoughts about how the digestive tract of annelids differs from that of turbellarian flatworms.

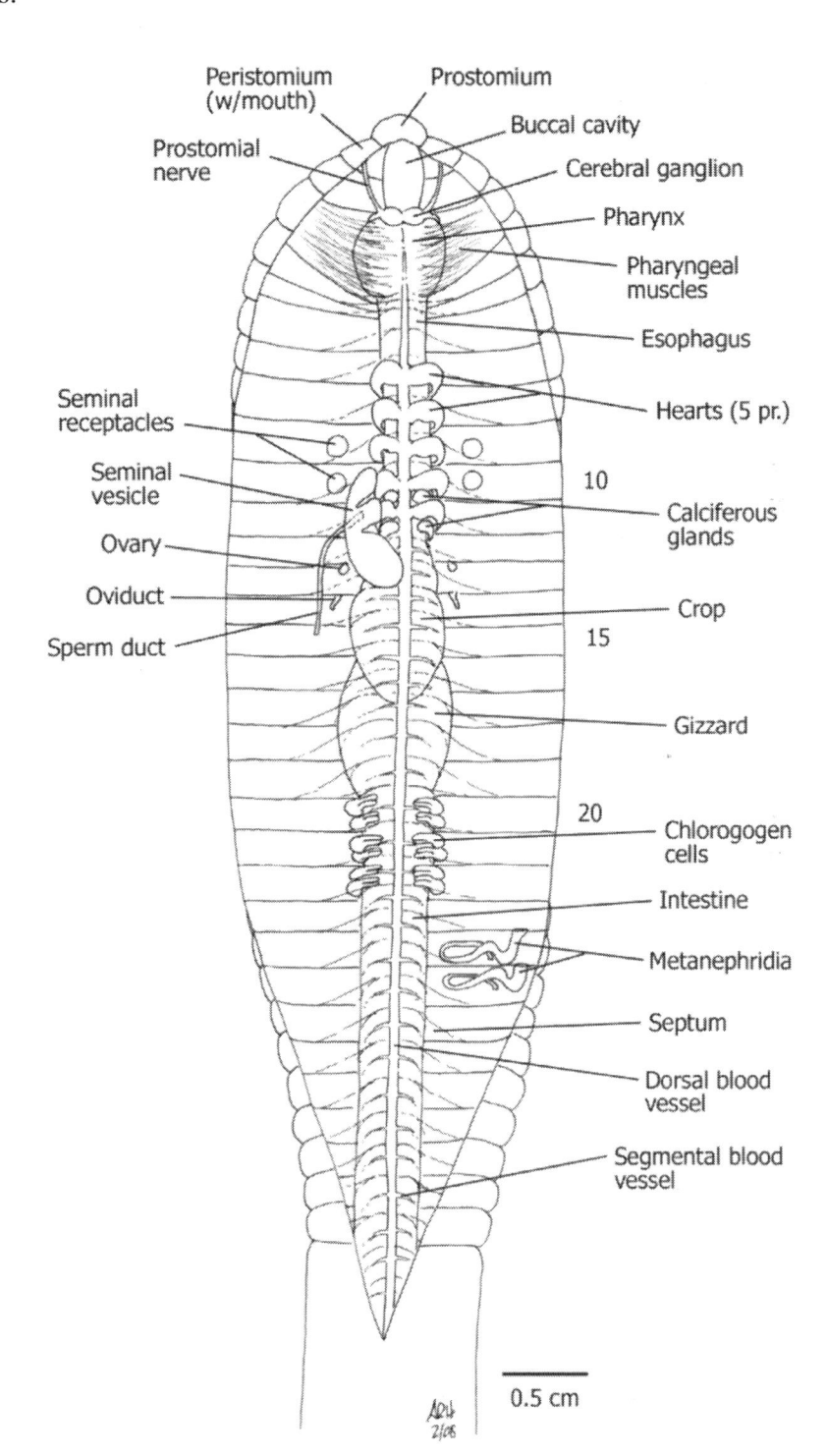

Figure 9.5. Internal anatomy of an earthworm. Be advised that some of the reproductive organs such as the ovaries, oviducts and perhaps the seminal receptacles will not be visible unless the animal is sexually mature and reproductive. (Image: ARH)

Phylum Mollusca (mollis = soft; or molluscus = soft shell)

There are about 100,000 described species of molluscs, making Phylum Mollusca the second largest animal phylum by number of described species. Only Phylum Arthropoda is larger.

Phylum Mollusca includes an interesting diversity of forms, and they make a living in every imaginable way: predators, herbivores, scavengers, suspension feeders, parasites and some even have symbiotic relationships with photosynthetic **zooxanthellae**. Molluscs have soft, unsegmented bodies that usually include a head, a muscular foot, **visceral mass**, specialized **mantle** tissue that secretes a calcified shell or shells, and a **mantle cavity**. Many molluscs also have a **radula** that they use during feeding.

There are eight classes of molluscs.

- **Class Caudofoveata** (caudo = tail, fovea = pit). Aplacophorans, also called spicule worms. These tiny marine worm-like molluscs secrete a chitinous cuticle and calcareous spicules but no shell. They lack a foot but have a pair of ctenidia in a small posterior mantle cavity and a radula is present.
- **Class Solenogastres** (soleno = pipe, gaster = belly). Aplacophorans, and like the Caudofoveata, also called spicule worms. These marine molluscs are also tiny, produce a chitinous cuticle and calcareous spicules as well as a foot groove but they lack **ctenidia**. The radula is lacking in some species.
- **Class Monoplacophora** (mono = one, placo = shell, phora = bearing) Monoplacophorans. These small snail-like, mostly deep sea molluscs produce a cap-shaped shell that looks like those of limpets. They have a circular foot with eight pairs of foot retractor muscles and three to six pairs of ctenidia located along both sides of the foot. Monoplacophorans are living fossils. Until the early 1950s, these animals were known only from fossils.
- **Class Polyplacophora** (poly = many, placo = shell, phora = bearing) Chitons. These marine molluscs have a dorsoventrally compressed body, with the dorsal surface protected by eight overlapping shell plates. The lateral edges of the plates are embedded in a girdle of tough, fleshy tissue. Shell plates of some species have organs in them called **aesthetes** that function as light sensory organs, like eye spots. Chitons have a broad muscular foot with long groove-like mantle cavities running along the length of the body between the foot and the girdle and contain up to 80 pairs of ctenidia. The head lacks tentacles but has a well-developed radula. Most chitons are herbivores that scrape algae or bacterial films off of rock surfaces or feed on fleshy algae. There is one predaceous species.
- **Class Gastropoda** (gastro = stomach, poda = foot) Snails and slugs. This is the most diverse group of molluscs with about 70,000 described species. Gastropods live in marine, freshwater and terrestrial environments. The diversity of their lifestyles is equally diverse and includes predators, herbivores, scavengers and even parasites. Members of this class carry out a developmental process called **torsion** during the larval stage. During torsion the visceral mass rotates 90-180°

relative to the head/foot. In slugs, torsion is followed by de-torsion. In snails, torsion results in the mantle cavity being located just above the head. This allows the head to be withdrawn quickly into the mantle cavity beneath the shell where it can be protected. Gastropods have a strong creeping foot, one or two pairs of cephalic tentacles, and a radula. Aquatic forms have ctenidia in the mantle cavity while terrestrial forms have a vascularized lung-like mantle cavity.

- **Class Bivalvia** (bi = two, valve = shell). Clams, mussels and relations. There are marine and freshwater species. Bivalves, commonly referred to as clams, have two calcareous shells connected to each other by a ligament. Large **adductor muscles** close the shells and the stiff proteinaceous **ligament** opposes these muscles so shells gape slightly when the adductor muscles relax. Clams lack a head and have a greatly enlarged mantle cavity that houses a pair of large ctenidia. The foot in most species is modified into a digging organ that is useful for burrowing in soft sediment. Most bivalves use large ctenidia for suspension feeding. Freshwater bivalves are particularly sensitive to poor water quality, making them important indicators of environmental quality.

- **Class Scaphopoda** (scapha = hollow, poda = foot). Tusk shells. These marine animals have a hollow, tusk-shaped shell up to 15 cm long with an opening at each end. The smaller end of the shell extends into the water above the soft substrate. Water is pulled in for respiration and pushed back out to release waste materials from the body. Like clams, tusk shells have a muscular foot used for burrowing in soft sediments. They also have many ciliated threadlike organs called captacula that are used to collect food particles from the sediment and transport them to the mouth.

- **Class Cephalopoda** (cephala = head, poda = foot). Octopus, squid and relations. Cephalopods are the most intelligent invertebrate animals. They have unusually large brains and exhibit memory, learning and problem-solving capabilities. They also have image-forming eyes. In some species, the mantle is modified into a muscular sac that contracts, squeezing out water forcefully through a tube called the siphon, producing jet propulsion. The foot is modified into prehensile arms and tentacles that bear many suckers or hooks. Cephalopods are predators that use a powerful jaw-like beak and toxins to kill prey and a radula to grind up food before it is swallowed. Most cephalopods have **chromatophores** in their skin that they can use for camouflage or communication. This video from Public Radio International's Science Friday program shows the camouflage capabilities of octopus: https://www.youtube.com/watch?v=eS-USrwuUfA

Task #4: The mollusc body plan

1) We don't know exactly what the ancestral mollusc looked like, but let's assume for the sake of this exercise that it looked like the hypothetical ancestral mollusc shown in **Fig. 9.6**. Obtain a preserved specimen of each of the following: snail, clam, chiton and squid. Examine each specimen and then describe evolutionary changes each it would have had to go through, starting with the hypothetical ancestral mollusc, to produce each body plan

you have in front of you. Make sure you include comments about each of the following for each of the molluscs you examine:
   a. Changes in the shape and location of the shell
   b. Changes in the shape and location of the visceral mass
   c. Changes in the shape and location of the head
   d. Changes in the shape and location of the foot
   e. Changes in the shape and location of the mantle cavity

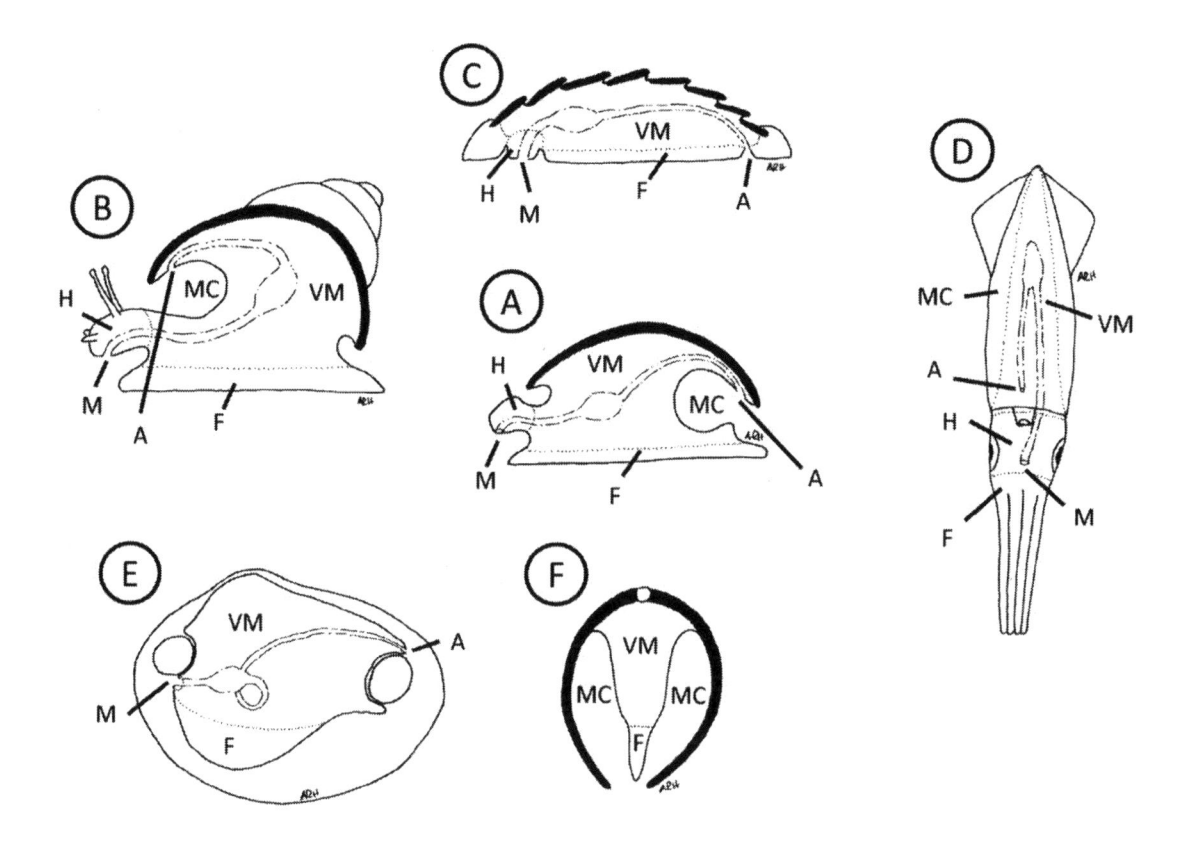

Figure 9.6. Mollusc body plans: **(A)** Hypothetical ancestral mollusc; **(B)** Gastropoda; **(C)** Polyplacophora; **(D)** Cephalopoda; **(E)** Bivalvia – internal side view; **(F)** Bivalvia – cross-section view. Major body regions (head, foot, visceral mass) are designated by stippled lines. Shells are indicated by heavy black lines. Note: The mantle cavity in polyplacophora **(C)** is located along each side of the foot between the head and the anus. Labeled structures: **A** - anus, **F** - foot, **H** - head, **M** - mouth, **MC** - mantle cavity, **VM** - visceral mass. (Image: ARH)

Group Questions

1) List advantages that a bilateral body plan has that a radial body plan does not.
2) Develop a hypothesis that could explain how a parasitic life cycle could have started in the first place.
3) Explain why you think annelid worms devote so much less space internally to reproductive organs than trematode flukes do.

4) Molluscs are the second largest phylum by number of species, and most of these are gastropods. Based on what you know about snails, develop a hypothesis that explains what makes the snail body plan so successful, and how it could give rise to aquatic and terrestrial forms.

Phylum Platyhelminthes – Glossary

Auricle – lateral extensions of the head that bear chemosensory cells

Cirrus – male copulatory organ of tapeworms

Cirrus pouch – cavity that houses the tapeworm cirrus

Diverticuli of the intestine – outpocketings or branches of the intestine that increase surface area and volume of the gut

Ectoparasite – parasite that attaches to the external surface of its host

Endoparasite – parasite that lives within the body of its host

Esophagus – tubular connection between the pharynx and the rest of the digestive tract

Excretory bladder – organ in flukes where nitrogenous waste is stored prior to being released

Eyespot – see Ocellus

Gastrovascular cavity – sac used for digestion and gas exchange, has only one opening that functions as both mouth and anus

Genital opening / genital pore – opening in flukes and tapeworms where sperm are transferred and embryonated eggs are released

Laurer's canal – is the tube through which sperm move to the seminal receptacle

Mehli's gland / shell gland – gland in flukes that produces the protective outer covering or shell for embryos, it also secretes mucus that provides lubrication that helps move encapsulated embryos through the uterus

Nephridiopore – opening through which nitrogenous waste is released from the body

Ocellus / eye – cup-shaped structure that detects the presence, intensity and direction of light, it may be able to detect movement but it does not generate an image

Oral sucker – muscular feeding and attachment organ in flukes, surrounds the mouth

Ovary – primary female reproductive organ, produces eggs

Pharyngeal cavity – space that houses the pharynx in free-living flatworms

Pharynx – muscular tube used to pull material into the gut

Proglottid – compartmentalized body region of tapeworms that produces sperm and embryonated eggs and carries out copulation

Scolex – organ bearing hooks and suckers, used by tapeworms to attach to host's intestine

Seminal receptacle – organ that stores sperm after copulation

Testis – primary male reproductive organ, produces sperm

Triploblastic – body plan that produces three embryonic tissue layers: ectoderm, endoderm and mesoderm

Uterus – organ that houses embryonated eggs until they are released

Vagina – opening and tube through which sperm are received during copulation

Vas deferens – tubule in flukes that is formed when the vas efferens fuse and carries sperm to the common genital pore

Vas efferens – duct in flukes leaving the testis that carries sperm to the vas deferens

Ventral sucker / acetabulum – organ of attachment flukes, is located on the ventral surface of the body and may have accessory spines

Vitellarium – tissue/organ that produces yolk

Phylum Annelida – Glossary

Calciferous gland – organ that releases excess $CaCO_3$ into the gut

Cerebral ganglion – ganglion located in the head, dorsal to the gut

Chaetae – epidermal bristles made of sclerotized (hardened) chitin

Chlorogogen cells – a mass of cells lining the digestive tract that functions like a liver, storing and releasing nutrients as needed

Crop – thin-walled organ where ingested material can be stored temporarily

Gizzard – muscular organ that can grind up or mix ingested material

Lophophore – crown of ciliated tentacles used for suspension feeding

Metamerism – repeated structures in the body, e.g., segments in annelids

Metanephridia – excretory organ found in annelids that have a cilia-lined entrance, a tube where modification of the urine takes place and a bladder where urine is stored until it is released

Peristomium – second segment of the annelid body, has a pair of coelomic spaces and houses the mouth and cerebral ganglia

Prostomium – first segment of the annelid body, bears sensory tentacles in some species and always contains only one coelomic space

Seminal receptacle – female organ that receives sperm during mating

Seminal vesicle – male organ that stores sperm until it is used during mating

Septum – wall of peritoneal tissue that separates neighboring segments in the annelid body

Triploblastic – body that produces embryonic ectoderm, endoderm and mesoderm

Trochophore larva – larva with a double-band of cilia around the equator and tufts at the anterior and posterior ends

Tubule of metanephridium – portion of a metanephridium where the urine is modified by secretion and reabsorption of materials

Phylum Mollusca – Glossary

Aesthetes – shell canals/light sensory structures in chitons

Adductor muscle – muscle used to close the shells of bivalves

Chromatophore – pigmented cells used by cephalopods to change color

Ctenidia – mollusc gills

Ligament – stiff, flexible proteinaceous tissue used to connect hard structures, such as two shells, together

Mantle - tissue layer that secretes the shell and lines the inner surfaces of shells or may be thick and muscular as in cephalopods

Mantle cavity – fluid-filled cavity that houses the ctenidia and other organs

Radula – tooth-bearing tongue-like strap used for scraping up or rasping on food items during feeding

Torsion – 90^0 to 180^0 rotation of the visceral mass and shell relative to the head-foot in gastropods, contractions of larval retractor muscles and differential tissue growth in the veliger stage results produce a U-shaped gut and twisting of longitudinal nerve cords that extend into the visceral mass, also moves the mantle cavity into an anterior position

Trochophore larva – larval stage produced by most mollusc taxa, has an oblong body with a dome-shaped upper portion, two bands of cilia running around the equator that are used for locomotion and feeding and tufts of cilia at the dorsal and ventral poles

Visceral mass – the soft non-muscular region of the mollusc body that houses the internal organs, is located dorsal to the head/foot and is covered and protected by the mantle and shell

Zooxanthellae – autotrophic, endosymbiotic dinoflagellates

Chapter 10: Animalia - Ecdysozoa

Clade Ecdysozoa (ecdyso = molting or stripping, zoa = animal) together with Clade Lophotrochozoa (Chapter 9) constitute Clade Protostomia. Ecdysozoans undergo a unique hormone-moderated process called **ecdysis**, commonly referred to as molting. The term molting is not sufficient to describe what these animals do because the way they shed their outer covering differs from the process of molting carried out by other animals, like birds that shed feathers or mammals that shed fur. During ecdysis a new cuticle or exoskeleton is secreted and then replaces the old one. This process is under the control of two hormones: **ecdysone** promotes ecdysis, and **MIH** (molt inhibiting hormone) inhibits ecdysis. The levels of these hormones in the body shift in relation to life stage and internal conditions. All animals in this clade undergo ecdysis at least once and some do so many times during their lives.

There are eight phyla in Clade Ecdysozoa, including Phylum Arthropoda – the largest animal phylum. Phyla in Clade Ecdysozoa are listed in **Table 10.1**.

Table 10.1. Phyla in Clade Ecdysozoa. Major taxa are indicated with an asterisk.

Phylum name	Common name or representative	Habitat and ecology	Image
Priapulida	Penis worms	Microscopic to macroscopic predators in marine mud and sediments	
Loricifera	Loriciferans	Microscopic members of marine sediments	
Kinorhyncha	Mud dragons	Microscopic members of marine sediments	
Nematoda*	Roundworms	Free-living and parasitic worms, found in all habitats	
Nematomorpha	Gordian or horsehair worms	Endoparasites of some insects	

Tardigrada	Water bears or moss piglets	Marine, freshwater, semi-terrestrial microherbivores	
Onychophora	Velvet worms	Nocturnal predators in tropical moist forests	
Arthropoda*	Insects, spiders, crabs, etc.	Every possible habitat and lifestyle	

Phylum Nematoda (nema = thread, oda = like)

Roundworms may be the most numerically abundant animals on Earth. We know quite a lot about medically important parasitic roundworms but we know precious little about the non-parasitic forms. This makes estimating their diversity difficult. Most nematodes live in the soil or other habitats where they largely do not affect human health.

Nematodes undergo ecdysis a few times during their early development. One of the key clues that roundworms belong to Clade Ecdysozoa was the discovery that, like arthropods and other ecdysozoans, they also use ecdysone and MIH to control molting. Molecular phylogenetic analysis confirmed that nematodes belong in Clade Ecdysozoa.

Nematodes are called roundworms because their bodies are round in cross-section. They are unsegmented, triploblastic worms that have a tough outer cuticle made of β-chitin, four blocks of longitudinal muscles in the body wall but no circular muscles, and a **blastocoelom** (pseudocoelom) that can be spacious in some species to highly reduced in others. Characteristics unique to nematodes are chemosensory organs called **amphids** found on the head and used for locating prey and **phasmids** found near the posterior end of the body used for identifying mates. Many nematodes also exhibit **eutely**, a trait in which they produce a predetermined number of cells for an organ or, in some species, for the entire body. Eutely at the whole-body level makes the roundworm *Caenorhabditis elegans* (*C. elegans*) particularly useful for doing research in developmental biology because it consistently produces a fixed number of cells in its entire body.

<u>**Task #1: Roundworm behavior**</u>

1) Obtain some live vinegar eels (*Tubatrix aceti*) and place them in a small glass bowl with enough liquid to allow them to move freely, or make a wet mount side and observe them using either a dissection or compound microscope. Keep in mind that roundworms have only longitudinal muscles in their body walls and that their body walls are flexible but non-elastic. WRITE observations of how vinegar eels move and compare their movement to that of the earthworms you may have observed in an earlier lab (Ch 9).

Task #2: Roundworm anatomy

1) External anatomy of *Ascaris*: Work with a lab partner during this portion of the lab and be sure to glove up because female *Ascaris lumbricoides* contain thousands of embryonated eggs that can still be viable even after being immersed in toxic materials for prolonged periods of time. *Ascaris lumbricoides*, the hog intestinal roundworm, is the largest intestinal roundworm parasite found in humans, reaching lengths of 30 cm. Obtain preserved specimens of a female and a male worm. Females are longer and larger than males and females are pointed at both ends. Males are smaller and the body forms a pronounced hook at the posterior end. DRAW female and male worms.
2) Use a compound scope to examine prepared cross-section slides of female and male *Ascaris*. DRAW what you see and use **Figs. 10.1-4** to help you identify what you see. Though you are not expected to dissect a roundworm, **Figs. 10.1** and **10.2** are provided to help you visualize the internal anatomy of these worms.

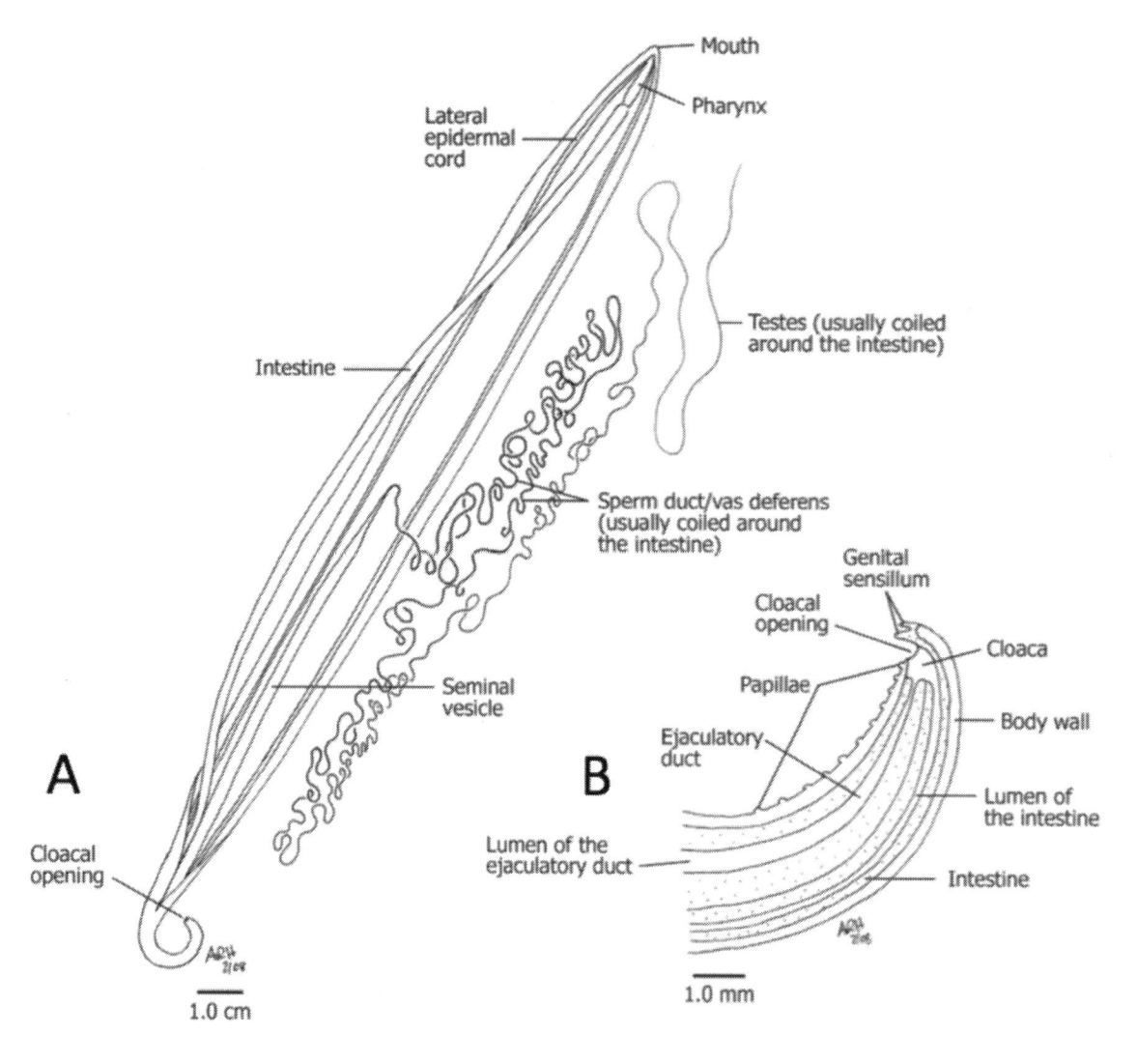

Figure 10.1. **(A)** Internal anatomy of male *Ascaris*, and **(B)** cross-section of the cloacal region. (Image: ARH)

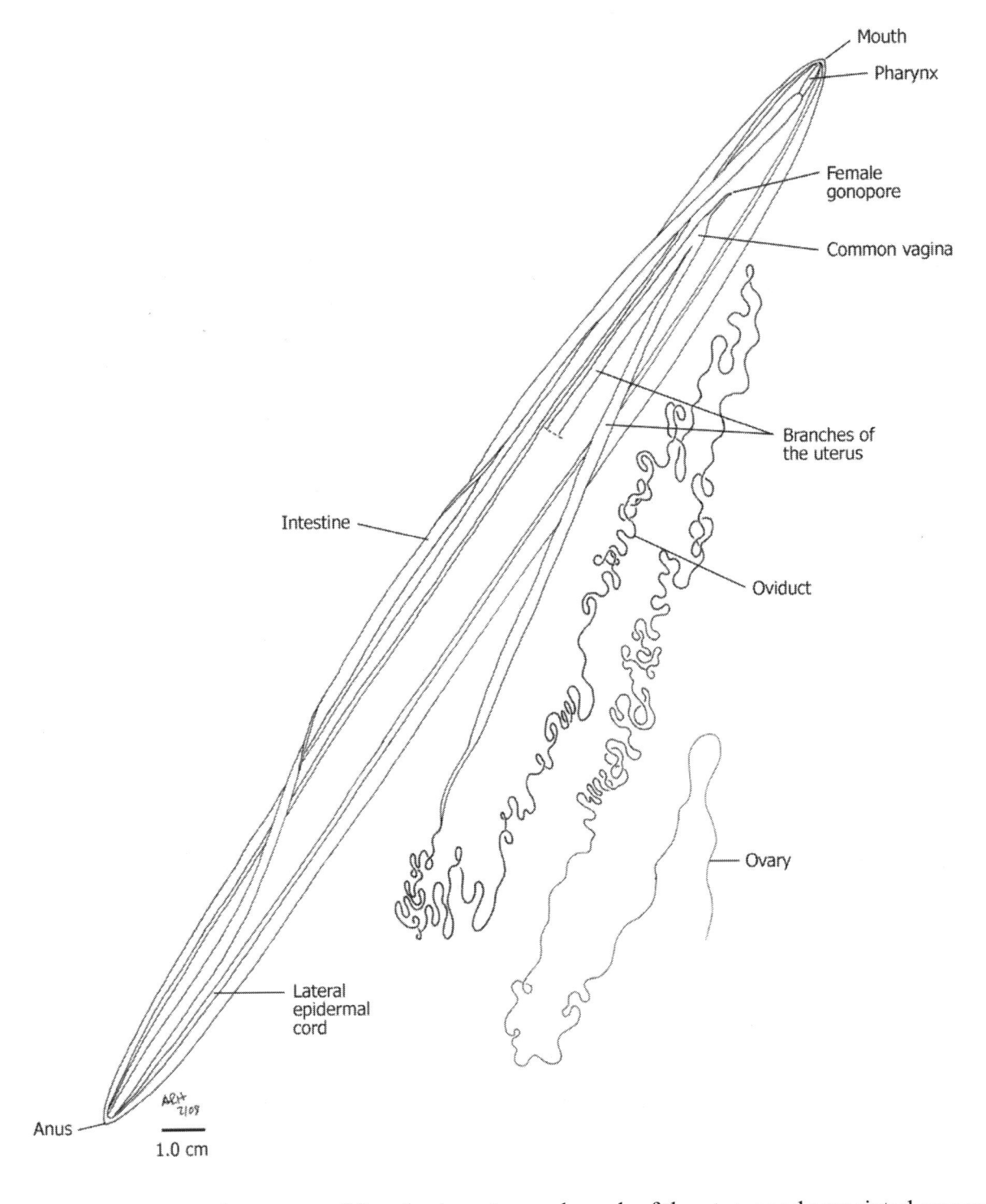

Figure 10.2. Internal anatomy of female *Ascaris*; one branch of the uterus and associated organs are not shown to simplify the image. (Image: ARH)

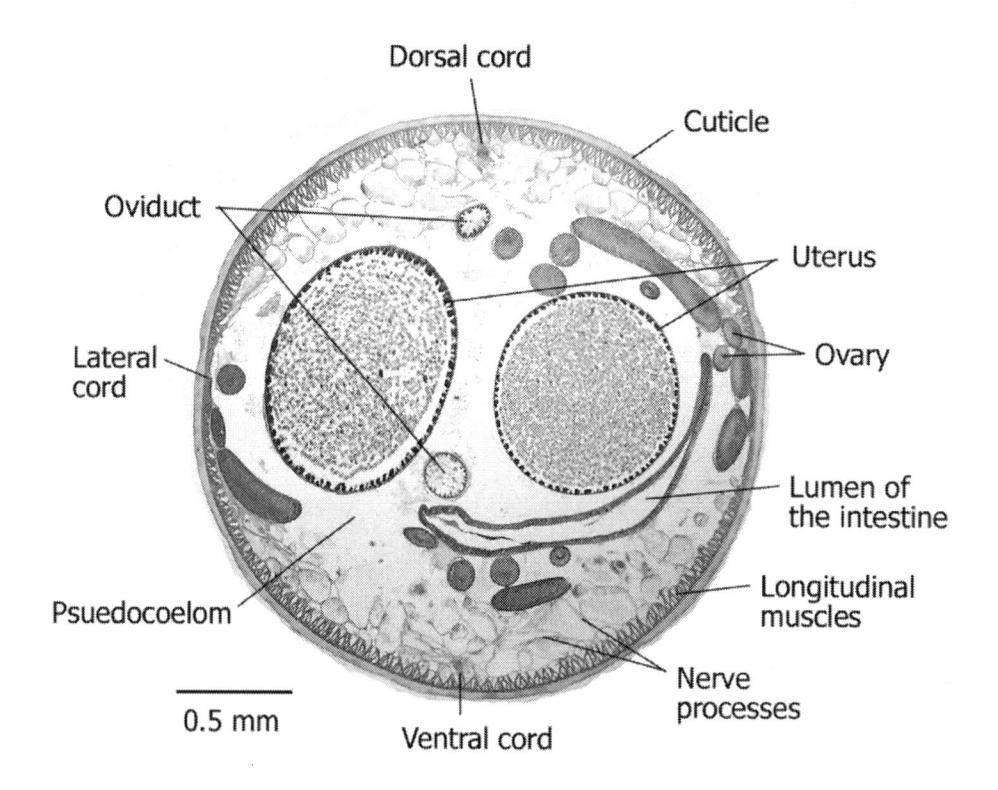

Figure 10.3. Cross-section through the intestine and uterus of female *Ascaris*. (Image: ARH)

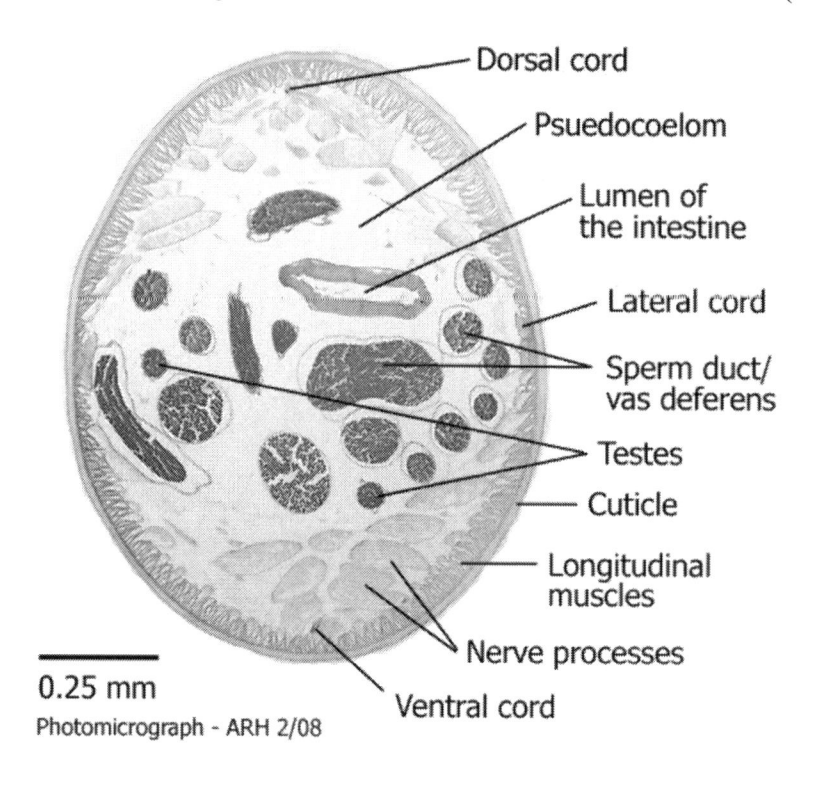

Figure 10.4. Cross-section through the intestine region of male *Ascaris*. (Image: ARH)

Task #3: Review life cycles of parasitic roundworms

1) *Ascaris lumbricoides* – hog intestinal roundworm (**Fig. 10.5**).

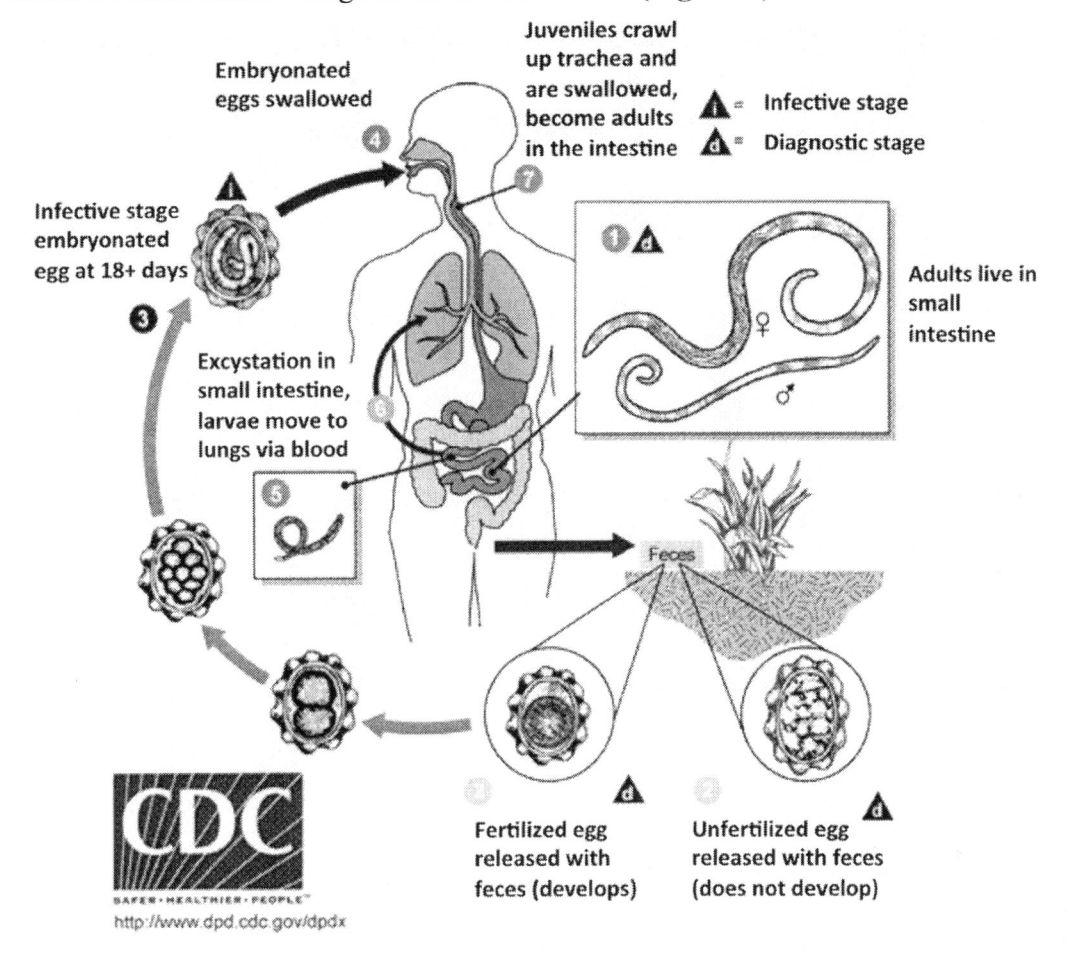

Figure 10.5. Life cycle of *Ascaris*. Adult worms live in the small intestine where they mate (1), eggs are released with feces and develop outside of the body (2-3), infective larvae are ingested (4) and excyst in the small intestine (5), they burrow through the intestinal mucosa and are carried by the blood to the lungs (6), they feed and grow in the lungs and when they approach maturity they crawl up the trachea to the pharynx where they are swallowed (7) and become adults in the small intestine. (Image: ARH modified public domain image from the CDC, http://www.cdc.gov/dpdx/ascariasis/)

2) *Trichinella spiralis* (**Fig. 10.6**). This common human parasite causes a condition called trichinosis. Trichinosis occurs when a person eats raw or improperly cooked meat, usually pork that contains infective stage larvae encysted in the flesh. Disease symptoms include abdominal pain and gastrointestinal problems within the first few days with chronic symptoms including muscle pain, intestinal problems, general weakness, fatigue and headaches thereafter. The latter symptoms are caused when next generation worms destroy host muscle tissue as they burrow in and encyst. This parasite is the reason that we are taught to cook pork thoroughly. The U.S. pork industry has nearly eradicated

trichinosis in pork grown indoors but hogs grown outdoors often have this parasite so keep on cooking those pork products!

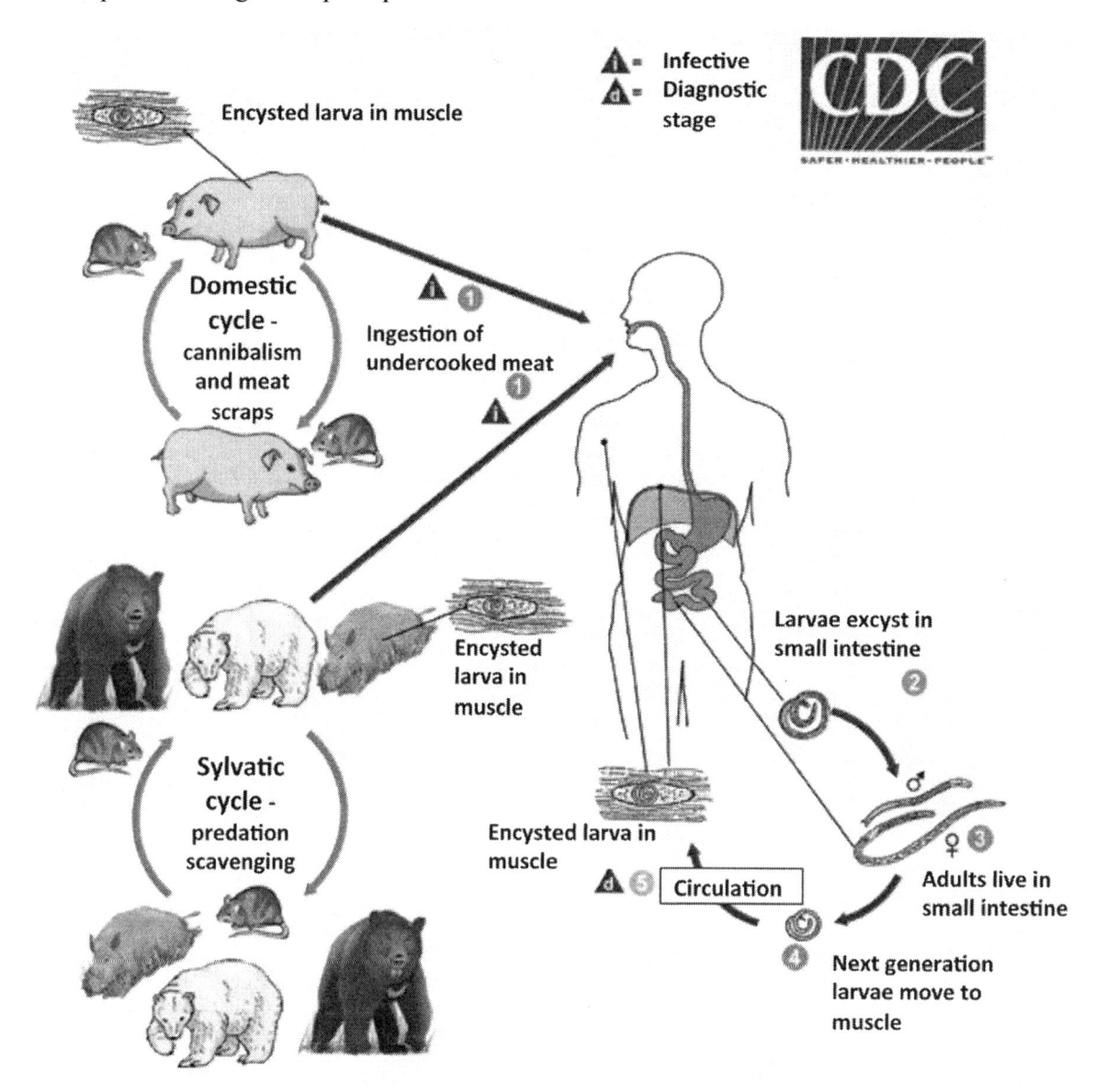

Figure 10.6. Life cycle of *Trichinella spiralis*. Infection in humans occurs when a person ingests raw or improperly cooked meat bearing encysted larvae (1). Larvae excyst, mature and mate in the small intestine (2-3). Next generation larvae burrow through the mucosa of the intestine (4) and into the bloodstream that carries them to skeletal muscle where they burrow in and encyst, destroying some muscle tissue as they do so (5). This life cycle is perpetuated among domestic and sylvatic (wild) animals through predator-prey or scavenging interactions in the wild and through cannibalism and ingestion of infected meat in domestic populations, mainly in populations of swine. (Image: ARH modified public domain image from the CDC, http://www.cdc.gov/parasites/trichinellosis/biology.html)

3) *Wuchereria bancrofti* (see **Fig. 10.**7). This filarial roundworm is a species that causes elephantiasis.

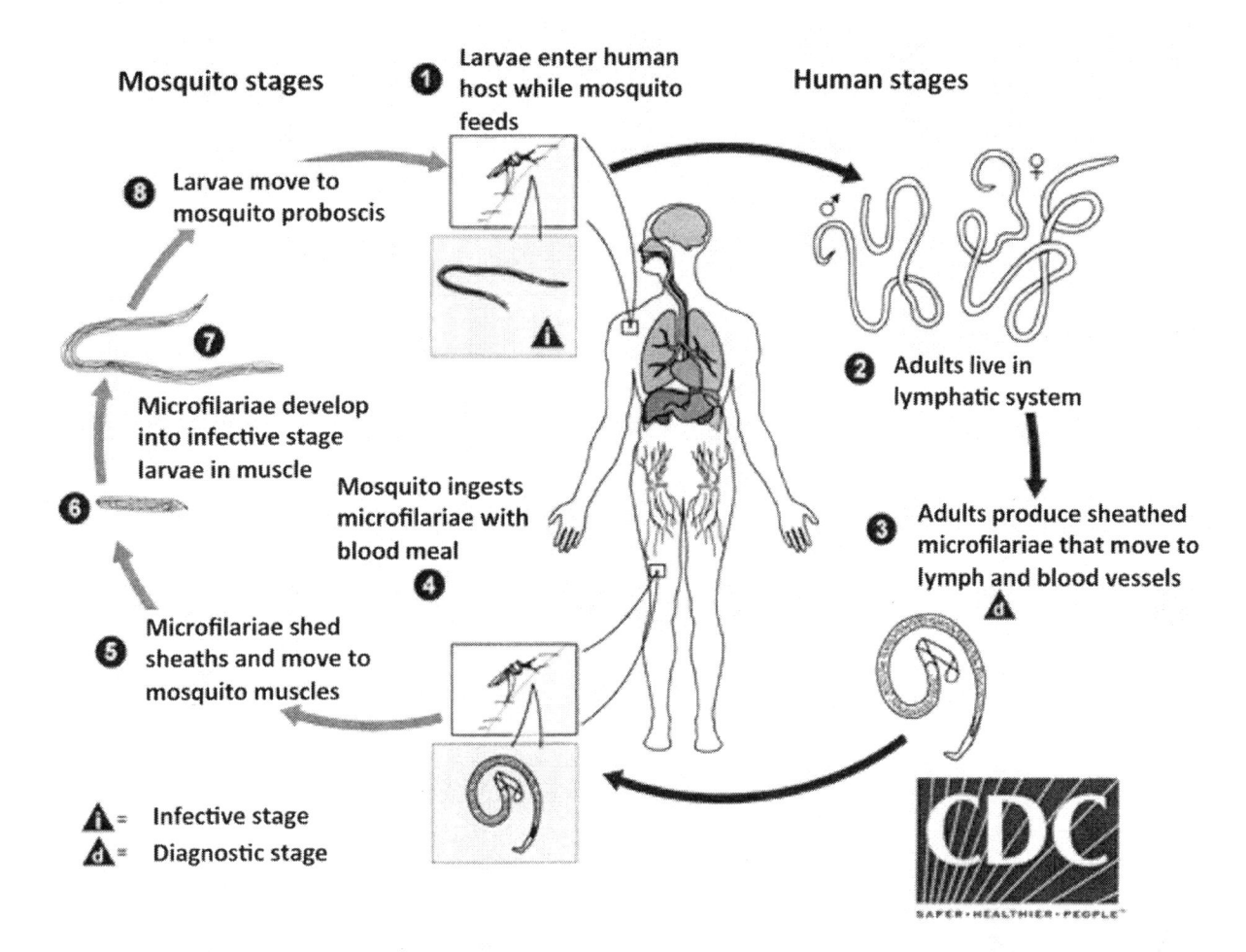

Figure 10.7. Life cycle of *Wuchereria bancrofti*. Infective larvae are injected into a human when a mosquito feeds (1). Larvae make their way to the lymphatic system where adults take up residence in the lymph nodes (2), adults clog up the lymph nodes causing the symptoms of elephantiasis. Females can reach 8-10 cm in length and males are about half that size. Adults produce microfilariae that get into the blood stream where they are ingested by mosquitoes (3-4). Microfilariae develop into infective stage larvae in the gut of mosquitoes and then migrate to the proboscis of mosquitoes (5-8). According to the CDC elephantiasis can be treated with a drug that kills microfilariae and some adult worms. (Image: ARH modified public domain image from the CDC, http://www.cdc.gov/parasites/lymphaticfilariasis/biology_w_bancrofti.html)

4) *Enterobius vermicularis* (see **Fig. 10.8**). Every prospective parent should know about this parasite. It is commonly known as pinworm and is the most common roundworm parasite in the USA.

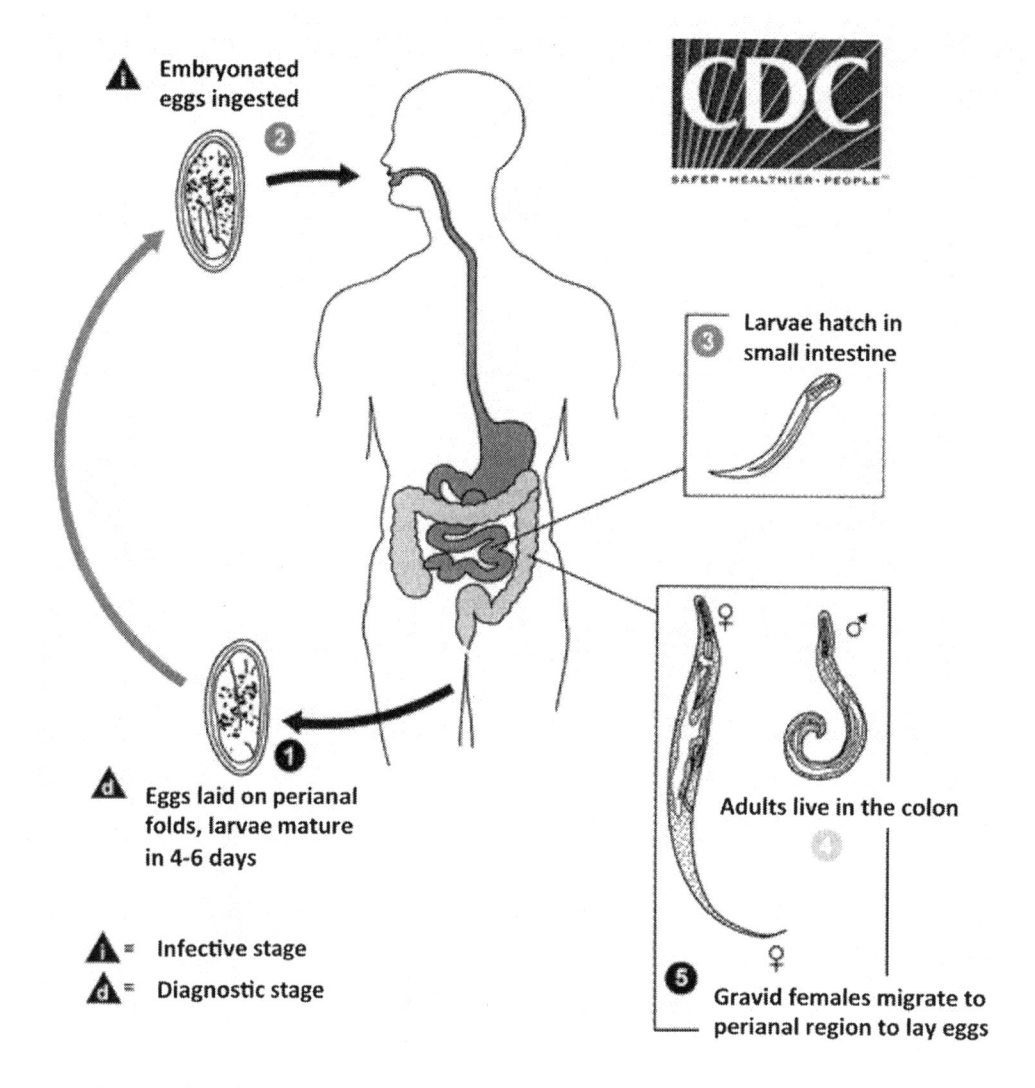

Figure 10.8. Life cycle of *Enterobius vermicularis*. This parasite is particularly common in children and among caregivers of children (especially those who bite their nails – think about it). Females emerge through the host's anus when the host is quiet, usually when the host is asleep. Females deposit **embryonated eggs** on the perianal region and return to the rectum through the anus (1), these larvae become infective in 4-6 hours and are incidentally ingested (2), they hatch in the small intestine (3) and become adults and live in the colon and **gravid** females emerge to deposit eggs as frequently as nightly (4-5). Eggs can cause itching in the perianal region and this can lead to direct reinfection. Eggs can also get onto clothing, bedding, etc., and lead to indirect infection. Heavy infection can produce chronic lower abdominal pain. The easiest way to check to see if someone (e.g., a child) has this parasite is to wait a few hours after they have gone to sleep, slip down their underclothes and use a flashlight to see if there are any worms around the perianal region. They will appear as moving whitish threadlike objects up around 1.0 cm long. An anti-helminthic drug treats this condition effectively. (Image: ARH modified public domain image from CDC, http://www.cdc.gov/parasites/pinworm/biology.html)

5) *Dracunculus medinensis* (see **Fig. 10.9**). The Guinea fireworm is on the verge of being eradicated globally, through efforts of the Carter Center – an initiative of former US President Jimmy Carter. If it is eradicated this will be only the second human disease to be completely eradicated. The first? Smallpox. The Guinea fireworm has a historic range covering parts of tropical Africa and Asia. Only a few cases per year are now being reported so eradication is within reach. You can learn more about this program and efforts to eradicate this roundworm parasite, a worm that has no treatment except slowly pulling the one-meter long worm out of the body a few centimeters at a time by watching this video: https://www.youtube.com/watch?v=u4kQWvUv_Ns.

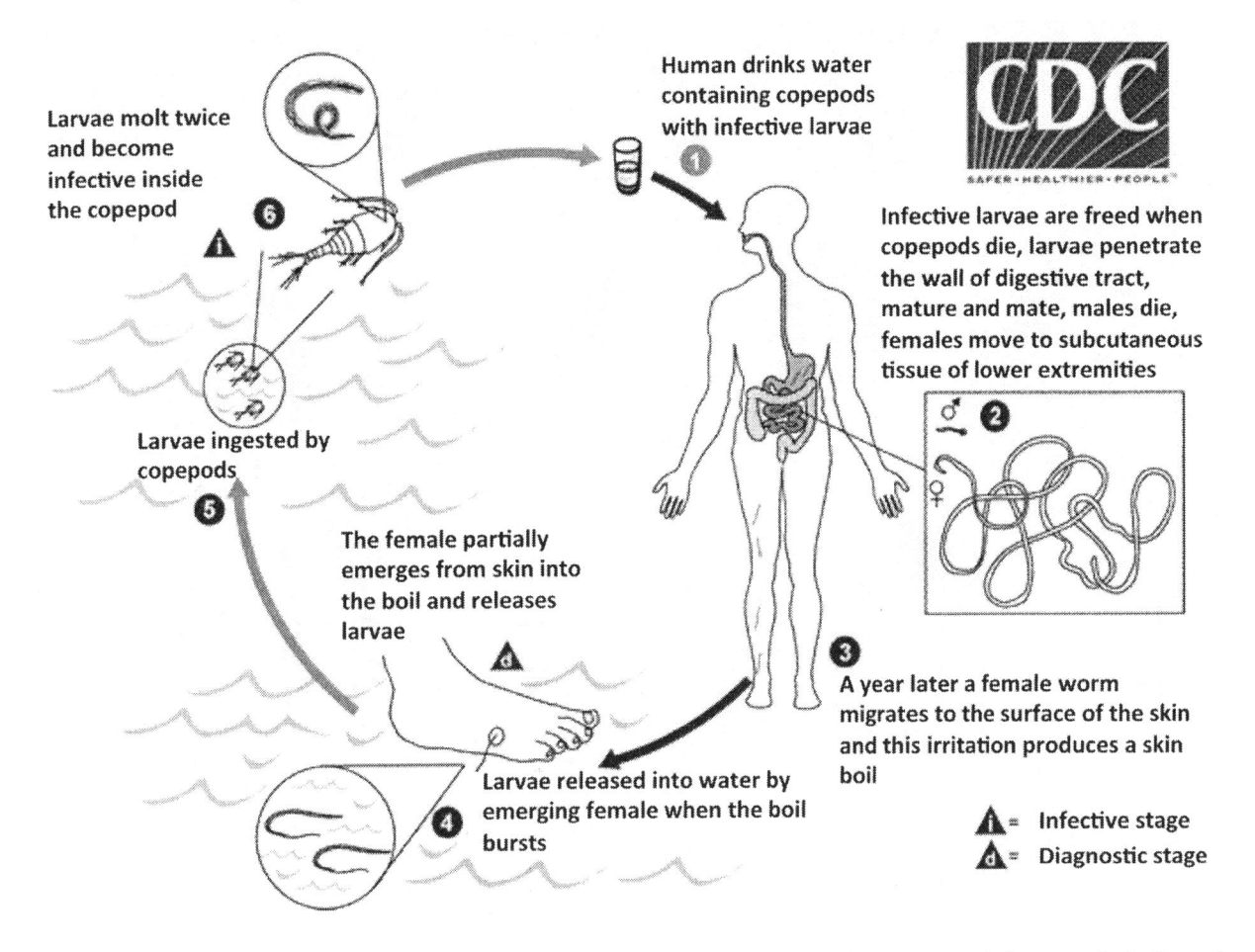

Figure 10.9. Life cycle of *Dracunculus medinensis,* the Guinea Fire Worm. A human is infected when they drink water that is contaminated with infected **copepods** (1). Larvae burrow through the human's stomach or intestine wall, matures and mates (2). Roughly a year later, females now up to one-meter long burrow into the subdermal tissues, usually of the lower extremities, and then into epidermis. The head of the worm irritates the skin and a fluid-filled blister or boil forms into which she extends the anterior part of her body (3). When the boil ruptures in water the female releases her larvae (4). These larvae are ingested by copepods. Ingested larvae develop into infective stage larvae inside their copepod hosts (5-6). Filtering or boiling drinking water effectively breaks this cycle. With any luck these worms will soon be extinct globally. (Image: ARH modified public domain image from the CDC, http://www.cdc.gov/parasites/images/guineaworm/dracunculiasis_lifecycle.gif)

6) *Dirofilaria immitis* – Dog heartworm (**Fig. 10.10**). Lest this lab exercise come across as too anthropocentric, it's only fair to include a parasite that affects other animals, in this case wolves and domesticated dogs, though humans are also hosts to these worms. There are three species of *Dirofilaria* that cause this disease but only *D. immitis* appears to affect domesticated dogs in North America. All of these are filarial roundworms that, like *Wuchereria*, move through the blood as microfilariae but in dogs the target tissue of adult worms are pulmonary arteries and sometimes the right ventricle of the heart, thus the common name of these worms. Fortunately, these worms do not develop into adults in humans.

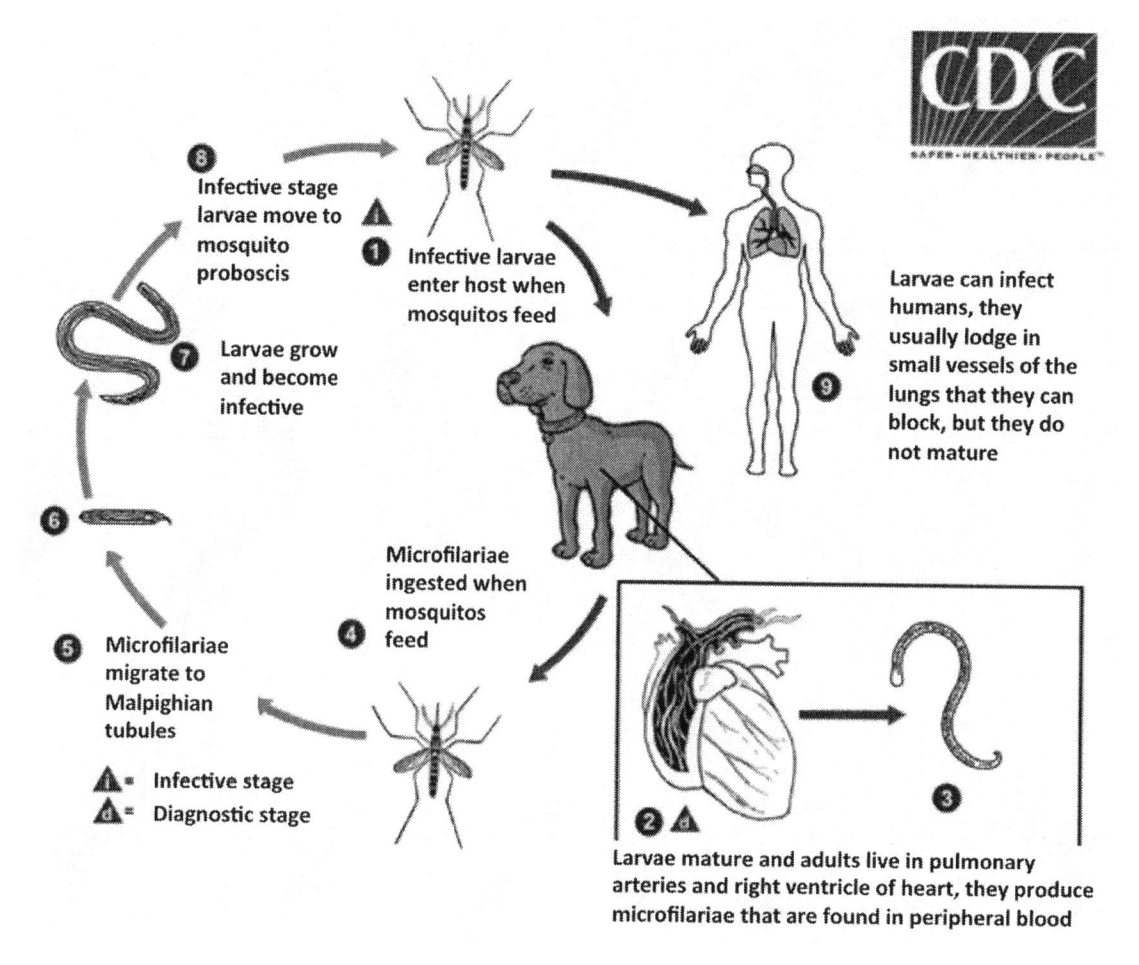

Figure 10.10. Life cycle of *Dirofilaria immitis*, the dog heartworm. Infective stage larvae are injected into a host by a mosquito during feeding (1) and larvae move to the pulmonary arteries and sometimes right ventricle in dogs where the worms take up residence (2), adults release **microfilariae** into the bloodstream (3) where they can be taken up by mosquitoes (4), and microfilariae develop into infective stage larvae in a mosquito's body, eventually making their way to the proboscis (5-8). Adult females reach lengths of 2-3 cm but males are smaller. According to the CDC adults can live up to 10 years with females producing microfilariae the entire time. Symptoms in dogs can include coughing up blood, blockage of vessels, fatigue and severe weight loss. Though these worms do not mature in humans, larvae migrate to blood vessels of the lungs where they can cause blockages and lesions. (Image: ARH modified public image from the CDC, http://www.cdc.gov/parasites/dirofilariasis/biology_d_immitis.html)

Clade Panarthropoda (pan = all, arthro = jointed, poda = foot)

Clade Panarthropoda is a taxon within Clade Ecdysozoa and includes the three phyla Tardigrada, Onychophora and Arthropoda. These phyla have a three-part brain, paired or fused ventral nerve cord, claws on the appendages, segmented bodies and they undergo ecdysis.

Phylum Tardigrada (tardi = slow, grada = to step)

There are about 1200 described species of tardigrades, commonly called water bears. Tardigrades go unseen and unappreciated because they are small, obscure and are apparently medically and ecologically unimportant. Most of these animals are only 0.1-0.5 mm long but giants can exceed 1.0 mm in length. Tardigrades are so tiny that many live within thin films of water on the surfaces of moist mosses and other plants as well as in aquatic environments. Tardigrades eat by sucking the cytoplasm out of plant cells after piercing them with a sharp **stylet**.

Tardigrades show evidence of segmentation and have four pairs of non-segmented, telescoping legs called **lobopods**. These animals undergo ecdysis periodically. They can also undergo **anabiosis** or **cryptobiosis** when environmental conditions deteriorate. During anabiosis a tardigrade will desiccate entirely and all biological functions cease. They can stay in this state of suspended animation apparently indefinitely. While in this condition tardigrades can survive just about anything including extreme temperatures, and, believe it or not, full exposure to the vacuum and other conditions of space.

<u>**Task #4:**</u> **Phylum Tardigrada**

1) Observe live tardigrades. Make a wet-mount slide. Be sure to put plasticine clay on the corners of your coverslip so that your specimens are not smashed. Use a pipette to collect samples from the bottom of the jar or bowl. Tardigrades will be the only things in your sample with four pairs of lobe-shaped appendages with chitinous hooks. When you find one, get excited, show your neighbor and then observe, describe and DRAW what you see. Be sure to write plenty of observations about their behavior. Refer to **Fig. 10.11** to help identify tardigrade anatomy.

Phylum Onychophora (onycho = nails or claws, phora = bearing)

Onychophorans are called velvet worms. It was once thought that this group of 200 species was evolutionary link between annelids and arthropods. Phylogenetic analysis reveals however that onychophorans are not a missing link but are instead the sister taxon to arthropods and are only very distantly related to annelid worms. Velvet worms have, like tardigrades, unjointed lobopods bearing hooks, as well as spiracles that cannot be closed. This explains why velvet worms are found only in the moist tropics – they would desiccate and die anywhere else.

This is also the only phylum of animals that is strictly terrestrial. There are no aquatic species or even aquatic life stages. They live in tropical moist forests where they are voracious nocturnal hunters. They capture their prey by spraying streams of sticky slime through their **oral/slime papillae**. The slime hardens as it is stretched when prey struggle to escape. They then slice through the body wall of their prey. Cool, but gruesome.

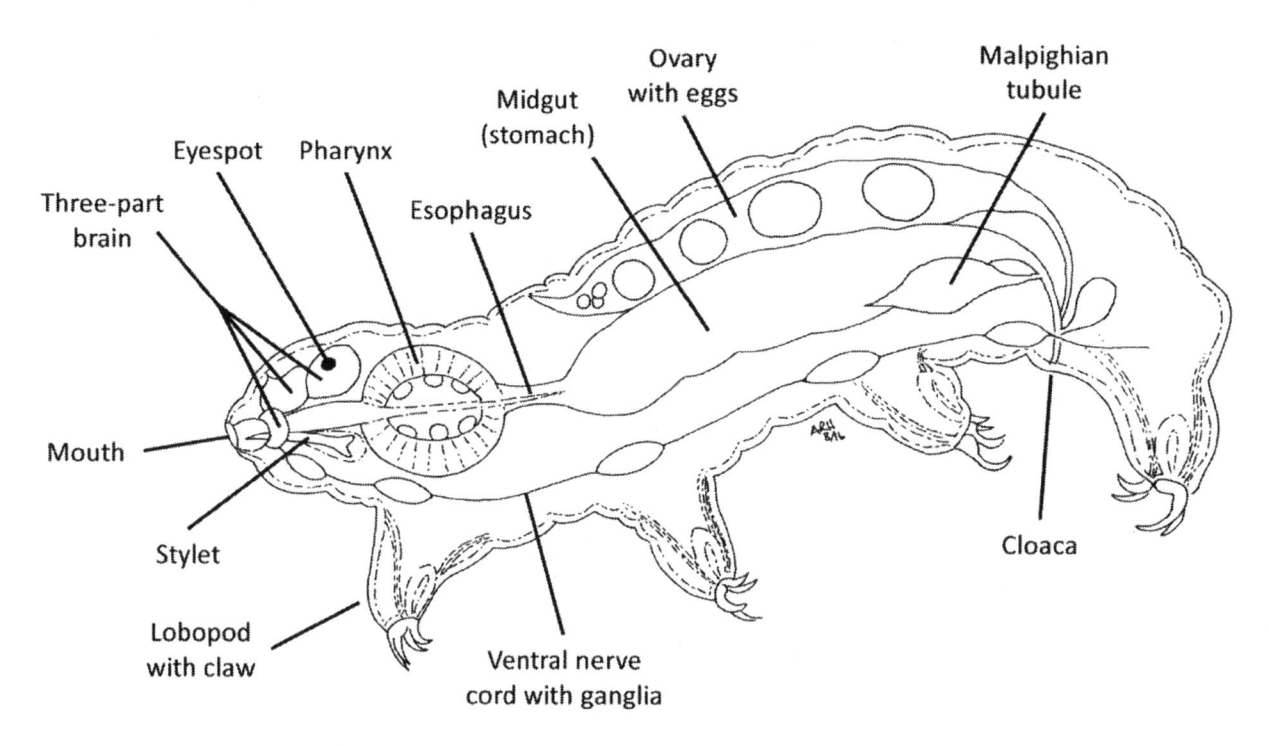

Figure 10.11. Tardigrade anatomy, only one set of appendages is shown. (Image: ARH)

Task #5: Phylum Onychophora

1) External anatomy of *Peripatus*. Use a magnifying glass or dissection scope to examine a preserved onychophoran. Observe, describe and DRAW the basic body plan. Use **Fig. 10.12** to help you identify what you see.

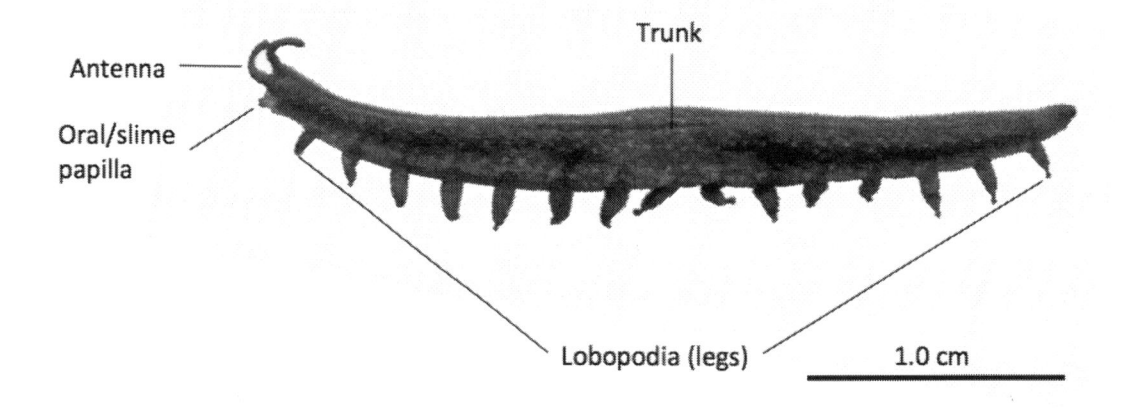

Figure 10.12. The onychophoran *Peripatus*. (Image: ARH)

Phylum Arthropoda (arthro = joint, poda = foot)

There are over one million described species of arthropods, making it the largest animal phylum. Arthropods occupy virtually every life-supporting habitat there is, including marine, freshwater and terrestrial environments, and employ every possible lifestyle.

The ancestral arthropod probably had a specialized head plus a trunk of many identical segments, each bearing a pair of identical appendages. This ancestral body plan evolved over time to produce the astounding diversity of arthropod body plans we see today.

Arthropods have a segmented body and a protective outer cuticle or exoskeleton that is molted periodically via ecdysis. Segments of the body are often fused together to produce specialized body parts called **tagma**, e.g., the head, thorax and abdomen of insects. Their jointed appendages are also specialized to carry out every kind of task imaginable, as depicted in these video clips from the on-line video series *The Shape of Life*: http://shapeoflife.org/video/marine-arthropods-successful-design, http://shapeoflife.org/video/terrestrial-arthropods-conquerors.

Phylum Arthropoda – Classification

Phylum Arthropoda includes five subphyla, four that are extant (living) and one that is extinct.

Subphylum Trilobitomorpha/Trilobita (tri = three, lob = lobes, morpha = form). Trilobites, extinct. Trilobites were prominent members of marine communities globally 500-225 million years ago and produced many index fossils – fossils useful in assigning ages to fossiliferous rocks. The last trilobites went extinct about 250 million years ago, during the Permian Extinction, also called The Great Dying, when 96% of all marine species went extinct.

Subphylum Chelicerata (chela = clawed, certata = horn). Horseshoe crabs, sea spiders, ticks, scorpions, spiders and relations. Most chelicerates have two body **tagma**: prosoma and opisthosoma. The prosoma bears a pair of **chelicerae**, **pedipalps** and four pairs of walking legs but lacks **antennae**. Chelicerates also have simple and **compound eyes**. Gas exchange organs vary between taxa but include **book gills**, **book lungs** and **trachea**.

Subphylum Myriapoda (myria = numberless, poda = foot). Centipedes, millipedes and relations. This body plan has a head and a trunk. The head bears one pair of the antennae and **mandibles**, plus two pairs of **maxillae**. The trunk is made of many identical segments that bear a pair of **uniramous appendages** and **spiracles**. These animals have only simple eyes.

Subphylum Crustacea (crusta = crust or hard). Shrimp, lobsters, crabs, barnacles and relations. This group contains ecologically and economically important species. They are found in marine, freshwater and terrestrial environments. There are crustaceans that are predators, herbivores, scavengers, suspension feeders and parasites. These animals usually have a **cephalothorax**, abdomen and two pairs of antennae, and aquatic taxa typically have a **nauplius** larval stage.

Subphylum Hexapoda (hexa = six, poda = foot). Insects and relations. Hexapods probably evolved within the Crustacea but for the time being they continue to be treated independently taxonomically. All hexapods have a head, thorax and abdomen. The head bears **antennae**, **mandibles**, **maxillae** and **labrum**. The thorax has three segments, each of which bears one pair of **uniramous legs**. Hexapods also have **compound eyes**, **spiracles**, **trachea** and **Malpighian tubules**. In addition, insects have two pairs of wings. Insects are the only group of animals that evolved wings without sacrificing a pair of walking appendages to make them.

<u>Task #6:</u> **Phylum Arthropoda – Subphylum Crustacea**

1) Crayfish. Observe, describe and DRAW the external anatomy of your specimen. Use **Fig. 10.13** to help you identify what you see.

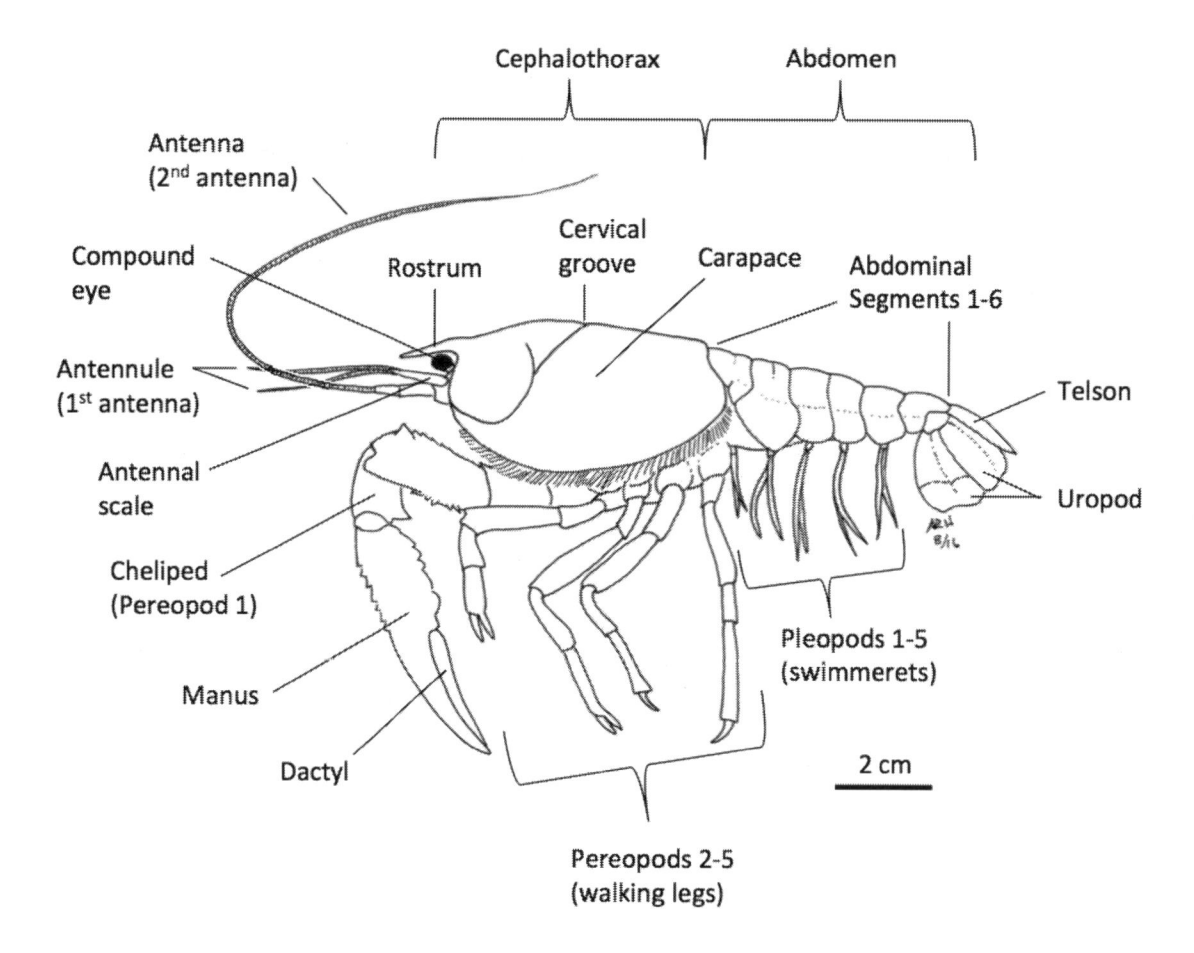

Figure 10.13. External anatomy of a female crayfish, lateral view. **Pereopods** 1-3 are **chelate appendages** (pincers) while **pereopods** 3 and 4 are **subchelate** – appendage tip flexes but does not pinch. (Image: ARH)

2) Crayfish appendages. Examine the diversity of appendages on the crayfish. Do not take the time to remove them, but look at them in place. **Fig. 10.14** shows all of the different appendages. Make a list of these appendages and do sufficient research to discover the function of each kind of appendage. Include that list in your lab manual.

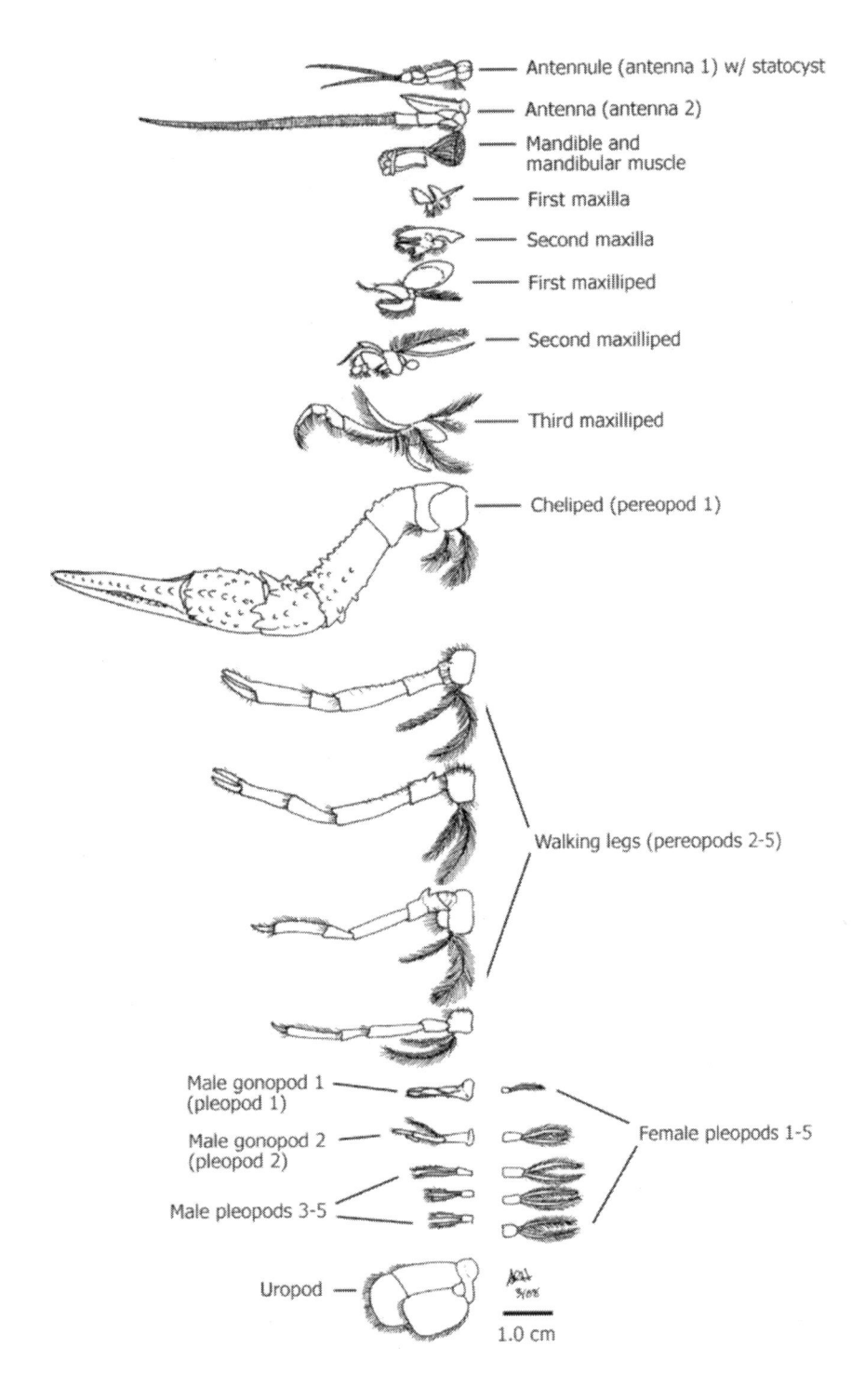

Figure 10.14. Appendages of crayfish in order, with anterior at the top to posterior at the bottom. Note the sexual dimorphism of pleopods. (Image: ARH)

3) Crayfish gills. Remove the carapace to expose the interior of the gill chamber.
   a. Use fine-tipped scissors make a cut through the exoskeleton of the cephalothorax, starting at the posterior edge of the carapace just off-center of the dorsal mid-line and continuing that cut as far anteriorly as possible.

b. Next, make a cut from the leading edge of the rostrum laterally until you reach the ventral edge of the carapace – do this on both sides of the specimen.
c. Insert a probe between the carapace and soft tissues of the body and separate any muscle attachments or other connections between soft tissues and the carapace. Carefully remove the carapace.
d. DRAW the gills in the gill chamber. Use the top and middle drawings in **Fig. 10.15** to help you identify what you see.

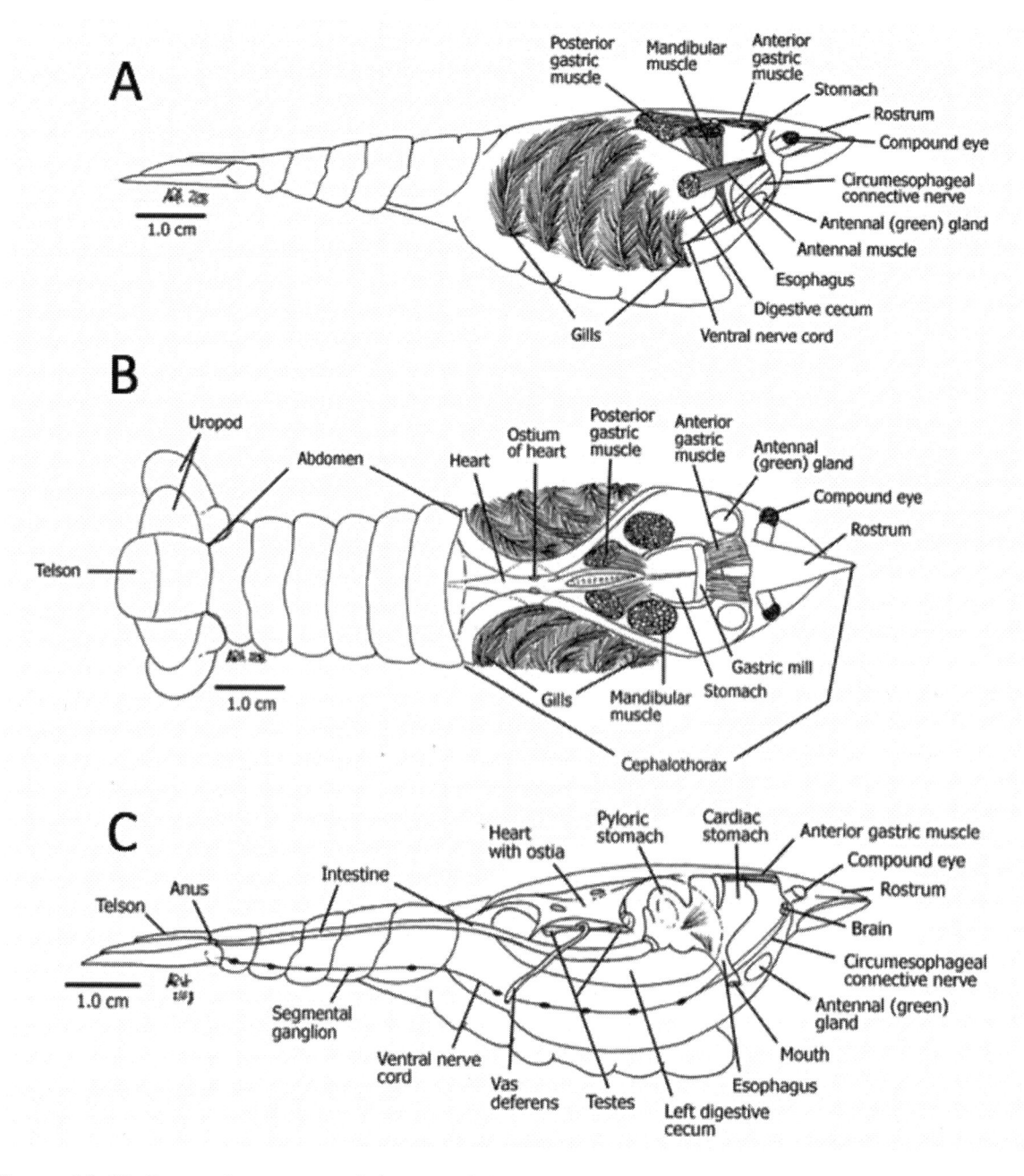

Figure 10.15. Internal anatomy of the crayfish. **(A)** Lateral view and **(B)** dorsal view of anatomy of the gill chamber and superficial soft tissues of the body cavity with the carapace removed, **(C)** lateral view of soft tissues visible when gills and digestive caecum are removed (male specimen) – appendages are not shown. (Image: ARH)

4) Crayfish internal organs.
   a. Remove the thin inner wall of the gill chamber and the large lateral digestive cecum in order to reveal other internal organs.
   b. Look for organs of the digestive tract and reproductive system.
   c. DRAW your specimen. Refer to the bottom drawing in **Fig. 10.15** to help you label your drawing.
5) Crayfish nervous system.
   a. Remove the muscles of the abdomen as well as any remaining organs in the **cephalothorax**.
   b. Look for the whitish ventral nerve cord and **segmental ganglia** that lie along the ventral wall of the exoskeleton.
   c. Look also in the rostrum to locate the brain.
   d. DRAW your specimen and refer again to **Fig. 10.15**, bottom drawing, for orientation to what you see.

Task #7: Phylum Arthropoda – Subphylum Hexapoda

1) Grasshopper external anatomy. Use a magnifying glass to look for spiracles on the abdomen, mouthparts and simple eyes. DRAW your specimen and use **Fig. 10.16** to help you identify what you see.

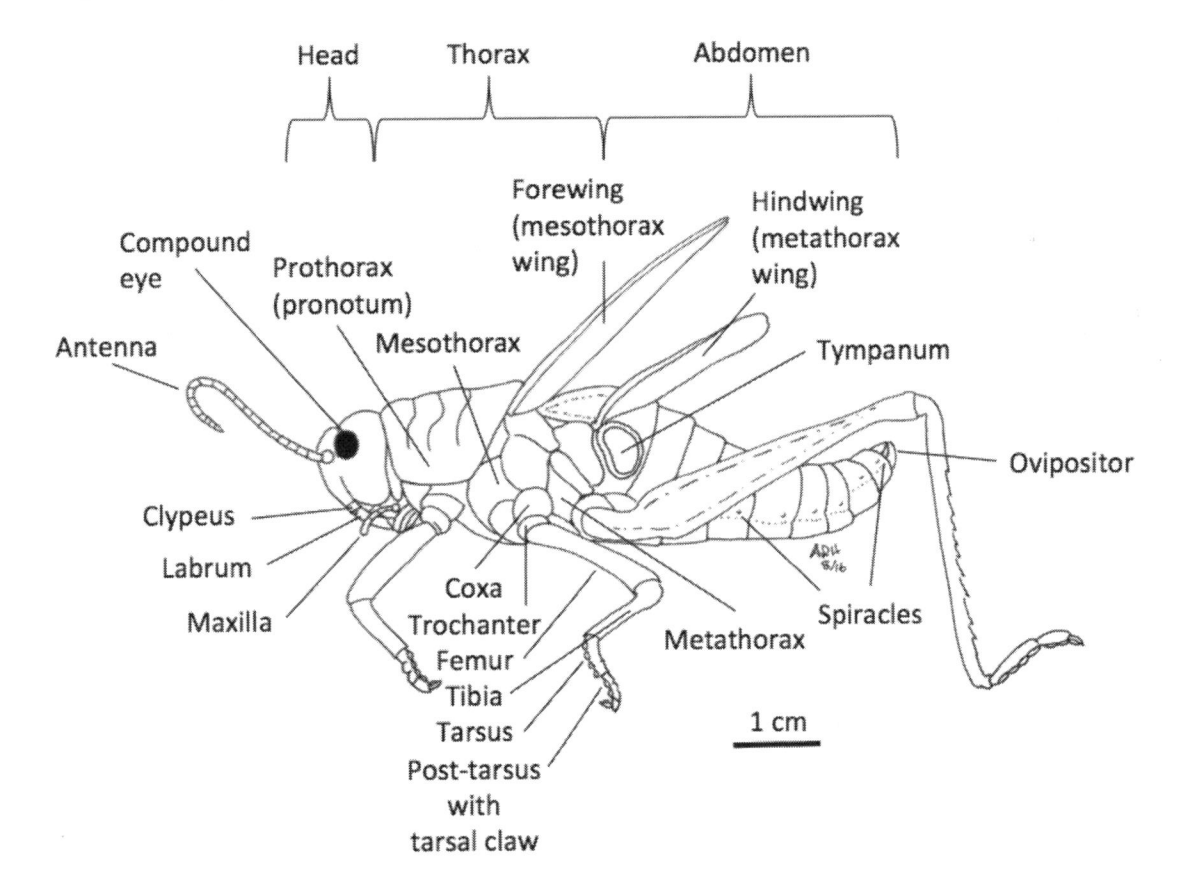

Figure 10.16. Anatomy of the lubber grasshopper. (Image: ARH)

2) Grasshopper Internal anatomy.
    a. Use scissors to remove the wings and walking legs (leave the coxa of all appendages in place).
    b. Use scissors to cut around the edges of the pronotum (see **Fig. 10.17**) and use a probe to scrape the inner surface of the pronotum as you remove it to detach any soft tissues attached to the inner surface of the carapace.
    c. Use fine-tipped scissors to make a longitudinal cut through the dorsal wall of the exoskeleton that runs the entire length of the thorax and abdomen.
    d. Make short lateral cuts along the exoskeleton and use insect pins to attach the exoskeleton to the floor of a wax-bottomed dissection pan. This gives you a dorsal view of the internal anatomy of your specimen.
    e. Immerse your specimen. You should see a thin layer of tissue (often reddish) covering much of the internal body cavity. This reddish tissue includes longitudinal and circular abdominal muscles.
    f. Look for the diamond shaped heart and dorsal vessel located dorsal to the midgut.
    g. After you have made a few observational notes use forceps (not scissors!) to carefully remove the reddish tissue layer. DRAW what you see. Refer to **Fig. 10.17** to help you identify structures of the internal anatomy of your grasshopper.

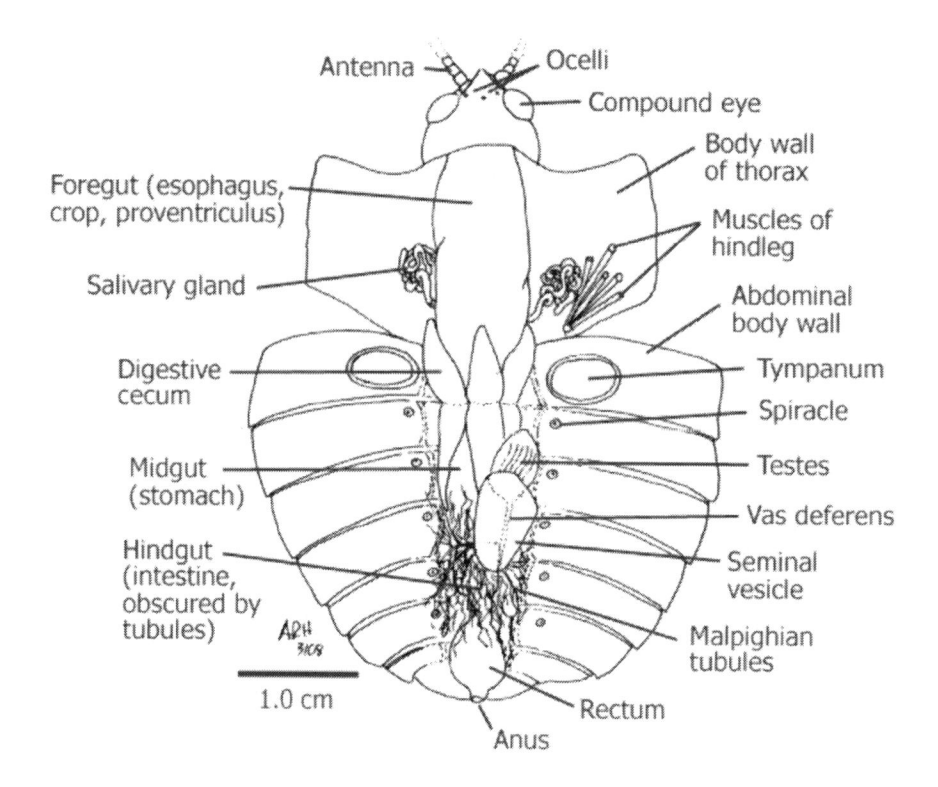

Figure 10.17. Dorsal view of the internal anatomy of a male lubber grasshopper with the dorsal vessel and heart removed. (Image: ARH)

3) Grasshopper nervous system.
    a. Remove all of the organs from the body cavity but do not disturb the thin layer of tissue lining the floor of the body cavity.

b. Immerse your specimen and use a magnifying lens or dissection scope to locate the whitish to translucent thread-like structures of the ventral nerve cord, segmental ganglia of the abdomen and many radiating nerves of the thoracic ganglia.
c. DRAW what you see.

Group Questions

1) Explain why animals as different as roundworms and crabs belong to Clade Ecdysozoa.
2) Based on what you learned in this lab, explain why roundworms are such effective endoparasites.
3) Describe one significant ecological cost and one significant ecological benefit of having an exoskeleton.
4) List features that make the arthropod body plan perhaps the most diverse and successful of all time.

Phylum Nematoda – Glossary

Blastocoelom – fluid-filled space derived from the embryonic blastocoel

Cuticle – protective outer covering secreted by the epidermis, made of β-chitin in nematodes

Dorsal cord – thickened area of the body wall that houses a dorsal nerve cord

Ecdysis – process of molting the cuticle

Ecdysone – hormone that promotes ecdysis

Embryonated egg – embryo encased in an environmentally resistant outer covering

Eutely – consistently producing a predetermined number of cells in an organ or the entire body

Gravid – full of eggs or embryonated eggs

Larva – post-hatching life stage that has a different appearance and ecology than adults and is not sexually mature

Lateral epidermal cord – thickened section of the body wall that houses the lateral excretory canals

Microfilaria – microscopic stage of nematodes taken up by an insect while feeding on blood

MIH – Molt inhibiting hormone, inhibits ecdysis

Ovary – primary female reproductive organ, produces eggs

Oviduct – tube that carries eggs from the ovary to the uterus

Pseudocoelom (see Blastocoelom)

Seminal vesicle – reservoir in the male body where sperm are stored in preparation for copulation

Sperm duct/vas deferens – tube between the testes and seminal vesicle

Uterus – female organ where fertilized and unfertilized eggs are stored until they are released

Ventral cord – thickened area of the body wall that houses a ventral nerve cord

Clade Panarthropoda – Glossary

Anabiosis/cryptobiosis – survival strategy used by tardigrades to survive difficult environmental stress, involves extreme desiccation and entering a state of suspended animation

Antenna – elongate segmented tactile and chemosensory structure (do not confuse antennae with tentacles that are non-segmented sensory structures, as in molluscs)

Antennule – first antenna of crustaceans, biramous tactile and chemosensory structure

Book gills – respiratory structure in some chelicerates with many sheet-like gills reminiscent of pages of a book that are protected by a hardened gill cover, e.g., horseshoe crabs

Book lungs – respiratory structures in some chelicerates that are sheet-like in structure and are completely enclosed in chambers accessible only via spiracles, e.g., spiders and scorpions

Carapace – shield-like exoskeletal structure that covers all or part of the dorsal surface

Cephalothorax – body tagma containing structures of the head and thorax

Chelicerae – first pair of appendages in chelicerates

Cheliped – appendage that bears an enlarged chelate or pinching structure, e.g., first pereopod of a crab

Cloaca – opening through which more than one organ system releases its product from the body, e.g., digestive system and reproductive system

Compound eye – optic organ made of many ommatidia (individual light collecting structures) that can produce high quality images

Cuticle – non-elastic chitinous outer body covering secreted by the epidermis

Dactyl – movable claw of a chelate appendage, word root means "finger"

Exoskeleton – hardened cuticle of arthropods, may be sclerotized (hardened) or calcified

Gills – gas exchange organs

Labrum – a flap-like structure of insects that covers the mandibles and produces a partially enclosed chamber where food can be manipulated and digestive enzymes can be extruded to begin digestion and food processing before being swallowed

Lobopodia – non-jointed legs of onychophorans and tardigrades

Malpighian tubule – thin tubular excretory organ that releases uric acid into the digestive tract at the junction of the midgut and hindgut

Mandibles – chewing or slicing mouthparts

Manus – rigid portion of chelate appendages, word root means "hand"

Maxillae – mouthparts used to taste and manipulate food

Maxilliped – appendage used as a fang in centipedes, used for food manipulation in crustaceans

Nauplius larva – first post-hatching larval stage of aquatic crustaceans, bears three pairs of appendages

Oral/slime papillae – a pair of knob-shaped structures on the head of onychophorans, used to shoot streams of sticky slime

Ostium of the heart – opening through which hemolymph is taken into the heart

Pedipalps – second pair of appendages in chelicerates, can be used for grasping, piercing, signaling, etc.

Pereopods – thoracic appendages of crustaceans, includes one pair of chelae (1st pair) and four pairs of walking legs in crayfish and crabs

Pharynx - muscular organ that pulls food into the digestive tract and in some animals also crushes it

Pleopods – abdominal appendages of crustaceans

Pronotum – exoskeletal plate that covers the anterior portion of the thorax in insects

Rostrum – portion of the crustacean carapace that extends farther forward than the eyes

Spiracle – opening through which air is pulled into trachea or a book lung of terrestrial panarthropods

Stylet – needle-like structure secreted by tardigrades that is used to pierce plant cells, tardigrades usually have two stylets at a time

Stylet gland – organ in tardigrades that secretes the stylet, a pair of these glands are located alongside the mouth

Tagma – multiple segments fuse together to produce specialized body parts, e.g., head, thorax or abdomen of insects

Telson – terminal body part extending beyond the last body segment in some arthropods, may be elongate as in horseshoe crabs or shorter as in crayfish, it never carries appendages and is not considered a true body segment

Thorax – middle body tagma, houses various internal organs including the stomach and heart and bears appendages

Tracheal system – respiratory system of branching tubules that carries air to individual cells

Tympanum – auditory (hearing) organ in insects

Uniramous appendage – jointed appendage that has no side branches

Uropod – flattened appendage attached to the last abdominal appendage in crayfish, lobster, shrimp, etc., used for rapid backward locomotion

Ventral nerve cord – solid cord of nervous tissue running along the ventral body wall

Chapter 11: Animalia – Deuterostomes

Clade Protostomia (Ch 9) and Clade Deuterostomia are taxa within Clade Bilateria, and some of the differences between them are listed in **Table 9.1**. There are three phyla in Clade Deuterostomia: Phylum Echinodermata, Phylum Hemichordata and Phylum Chordata (**Table 11.1**)

Table 11.1. Phyla in Clade Deuterostomia. Major taxa are indicated with an asterisk.

Phylum name	Common name or representative	Habitat and ecology	Image
Echinodermata*	Seastars, sea urchins and relations	Marine herbivores, carnivores, deposit, and suspension feeders	
Hemichordata	Acorn worms and pterobranchs	Marine deposit feeders and suspension feeders	
Chordata*	Lancelets, sea squirts, vertebrates	Marine, freshwater and terrestrial, diverse lifestyles	

Phylum Echinodermata (echino = spiny, derm = skin)

Phylum Echinodermata is a strictly marine taxon of about 7,000 species divided into five classes. Echinoderms are bilaterally symmetrical animals with secondary **pentaradial symmetry**. That is, as larvae they are clearly bilateral, but during metamorphosis they produce a body that usually has five or sets of five structures arranged radially around a central axis, though variations on this number exist. All animals in this phylum have a unique **water vascular system** that they use for locomotion, feeding or both. They also have a decentralized central nervous system, **mutable connective tissue** and a body wall that contains either calcareous plates or ossicles.

This video from the *Shape of Life* series introduces the body plan and diversity of echinoderms: http://shapeoflife.org/video/echinoderms-ultimate-animal

Phylum Echinodermata – Classification

Class Crinoidea (crino = lily, oid = like) Crinoids, feather stars and sea lilies, just over 600 living species. The oral surface faces away from the substrate and bears a centrally located, upward facing mouth surrounded by arms that extend outward from the central disc. A stalk supporting the body, when present, is attached to the aboral surface. Arms bear feather-like pinnules that are used for suspension feeding.

Class Asteroidea (aster = star, oid = like) Seastars, about 2000 species. Seastars have a **stellate** body with a central mouth surrounded by five or more unbranched arms. The oral surface faces the substrate and bears the mouth and **ambulacral grooves**. There is no clear demarcation between the arms and central disc in these animals. **Tube feet** of the water vascular system may or may not have suckers. The **madreporite** is on the **aboral surface**. The anus is often reduced and is also on the aboral surface.

Class Ophiuroidea (ophio = snake, oid = like) Brittle stars and basket stars, about 2000 species. These cryptic animals keep most of the body hidden and extend only an arm or two into the water and wave them – arm movement is reminiscent of the undulation of a snake. Ophiuroids have stellate bodies but the arms may be unbranched (brittle stars) or branched (basket stars) and the boundary between the arms and central disc is clearly visible. The body is covered by skeletal plates. Podia are peg-shaped and lack internal **ampullae**. The madreporite is on the oral surface and there is no anus. Arms of these animals readily break off when they are disturbed, thus giving brittle stars their common name.

Class Echinoidea (echino = spiny, oid = like) Sea urchins, sand dollars and relations. This group of about 1000 species has calcareous plates that are fused together to produce a rigid endoskeleton called a **test**. Moveable spines cover the body and they can use podia for locomotion, respiration or both. Some of these animals also have a complex chewing organ called **Aristotle's Lantern**.

Class Holothuroidea (holothur = a kind of animal that looks like a plant, oid = like), Sea cucumbers, about 1700 species. This includes animals with large worm-like bodies, a mouth is located at one end and the anus at the other. These animals lack a rigid test, but have small calcareous ossicles are embedded in the fleshy, muscular body wall. Pentaradial symmetry is expressed in the tentacles surrounding the mouth. The central axis is elongated with the mouth at one end and the anus at the other. Tentacles may be large and frilly and when extended may be reminiscent of a plant, thus the class name.

Class Asteroidea

Seastars are excellent representatives of Phylum Echinodermata. They are important members of marine environments where they are often top predators. Seastars have a global distribution and are members of intertidal and subtidal communities globally. Most people are familiar with seastars because they can be colorful and are easy to spot during low tide and while swimming or snorkeling. Plus, seastars are among the most iconic of marine invertebrate animals.

Task #1: Seastar anatomy

1) External anatomy of the seastar – aboral surface. Obtain a preserved seastar and examine its aboral surface. Locate the madreporite, the small button-shaped structure located on the aboral surface near the junction of two of the arms. These two arms are referred to as

the **bivium**. The other three arms are the **trivium**. DRAW the aboral surface of your specimen.
2) External anatomy of the seastar – oral surface. Look at the oral surface of your seastar. Identify the mouth and oral spines, the arms and ambulacral grooves that run along the oral surface of each arm. Look carefully at an ambulacral groove and identify spines and podia - the portion of tube feet that extend beyond the body wall. DRAW what you see.
3) Internal anatomy.
    a. Use a pair of scissors to make a shallow cut around the aboral edges of the **central disc**, but cut around the madreporite thus leaving it intact and in place. Use a probe to separate any soft tissues from the body wall as you carefully lift the body wall off of the central disc.
    b. Next, cut away the aboral body wall from the two arms of the trivium of your specimen. Identify structures of the digestive system in the central disc as well as organs housed in the arms. Refer to **Figure 11.1** and identify the structures on your specimen. DRAW what you see.

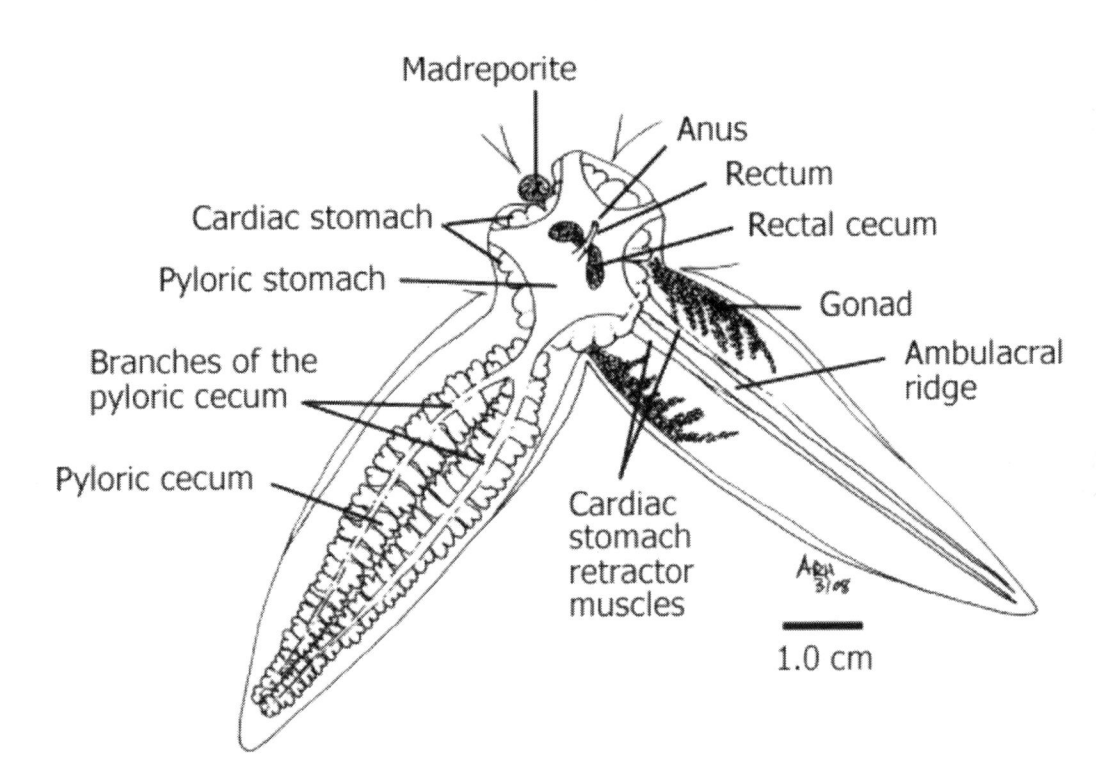

Figure 11.1. Internal anatomy of the central disc and arms of a seastar, aboral view. The arm to the right has had the pyloric cecum removed to uncover the gonads. (Image: ARH)

   c. Remove the **pyloric cecum** from one of the arms of your specimen. Locate the pair of gonads that are situated beneath the pyloric cecum. The gonads may be large or very small depending on the reproductive condition of your specimen.
   d. Also, identify the **cardiac stomach retractor muscles** by pulling gently on the **cardiac stomach** and looking for the nearly transparent cardiac stomach retractor muscles that are attached to the ambulacral ridge. Ponder what these muscles do and WRITE your conclusions.

4) Water vascular system.
    a. Remove the **pyloric** and **cardiac stomachs** from the central disc after you have identified them. Be sure to keep the madreporite and **stone canal** in place and intact. DRAW the entire water vascular system. Refer to **Figure 11.2** to help you identify what you see.

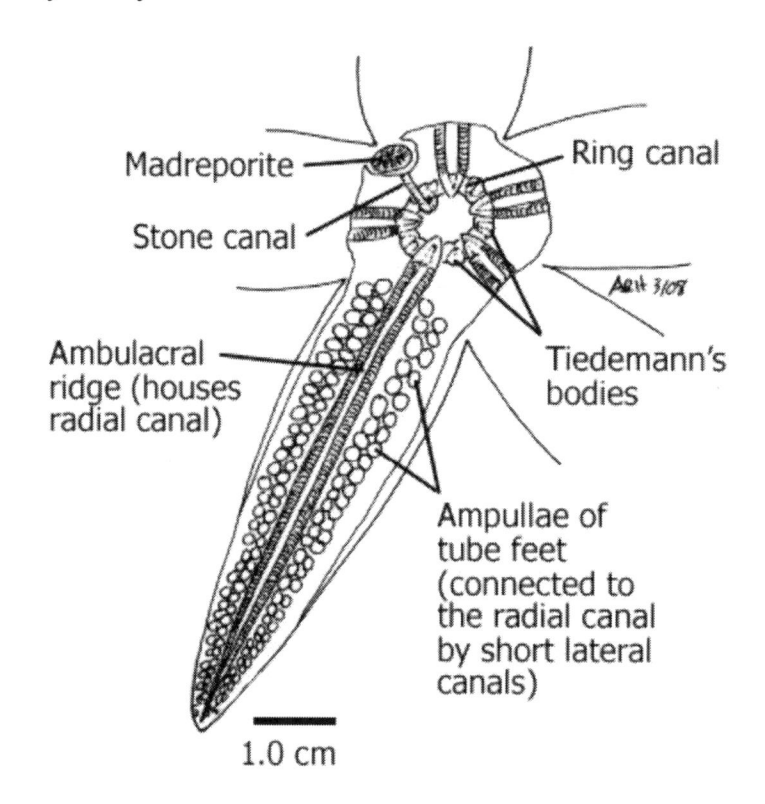

Figure 11.2. The water vascular system of a seastar, aboral view, with internal organs removed. (Image: ARH)

Phylum Chordata (chord = a string)

Phylum Chordata includes three subphyla: Cephalochordata, Urochordata and Vertebrata. The vast majority of chordates, including humans, belong to Subphylum Vertebrata, animals that produce a backbone. All chordates have **pharyngeal gill openings**, a dorsal **notochord**, a **dorsal hollow nerve chord**, a muscular **post-anal tail** and an **endostyle/thyroid** gland.

Watch this segment from *The Shape of Life* video series to see an introduction to the chordate body plan and to see how animals as diverse as sea squirts and humans are members of the same phylum: http://shapeoflife.org/video/chordates-we're-all-family

Phylum Chordata – Classification

Subphylum Cephalochordata (cephala = head, chord = string) Lancelets, about 30 species. Lancelets are small elongate chordates that lack vertebrae and a cranium. These strictly marine organisms make their living as **infaunal** suspension feeders. They live in shallow water in clean sand habitats where they burrow tail-first into the sediment. Only their heads protrude above the sediment surface so they can pull water into their mouths as they filter-feed. Lancelets are weak swimmers that tend to stay in one place.

Subphylum Urochordata (uro – tail, chord = string) Sea squirts and relations, about 3,000 species. Urochordates are also strictly marine. They are called urochordates because all the defining characteristics of Phylum Chordata are seen only in the tail-bearing, swimming tadpole larval stage. Most urochordates are suspension feeders that pull water through the pharynx and capture small particles on sheets of mucus secreted by the endostyle.

Subphylum Vertebrata (vertebr = joint) Animals with a cranium and backbone, about 58,000 species. These animals are covered in Chapter 12.

Subphylum Cephalochordata

Cephalochordates were once considered to be the sister taxon of the vertebrates but phylogenetic analysis definitively shows that they are not, urochordates are. Cephalochordates are therefore the sister taxon to the clade containing urochordates and vertebrates and has the basic body plan from which all chordates arose.

Task #2: Cephalochordate body plan

1) Obtain a preserved lancelet. Use a hand lens or dissection scope to look for metameric chevron-shaped muscle blocks that line the flanks of these animals. Also, look for many cream-colored gonads located along the ventral half of the body.
2) Obtain a prepared whole mount slide of a cephalochordate and study its internal anatomy. DRAW what you see. Refer to **Figure 11.3** to help you identify the anatomy of your specimen. Take particular care to identify the anatomical structures that qualify this animal to be a member of Phylum Chordata.

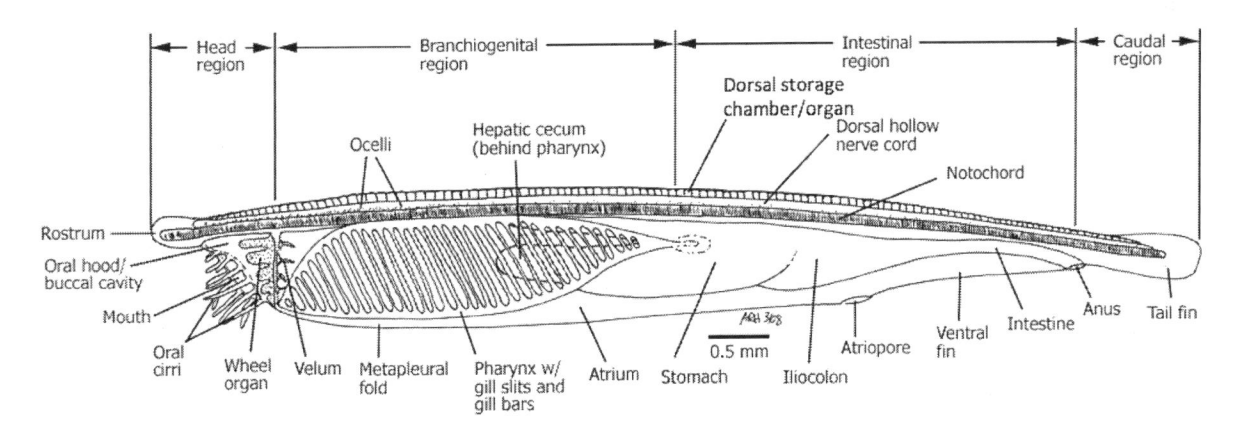

Figure 11.3. Lateral view of a cephalochordate. (Image: ARH)

3) Obtain a prepared cross-section slide through the pharyngeal region of a cephalochordate. Use a compound scope to study the slide and DRAW what you see. Use **Figure 11.4** to help you identify the anatomy of your specimen.

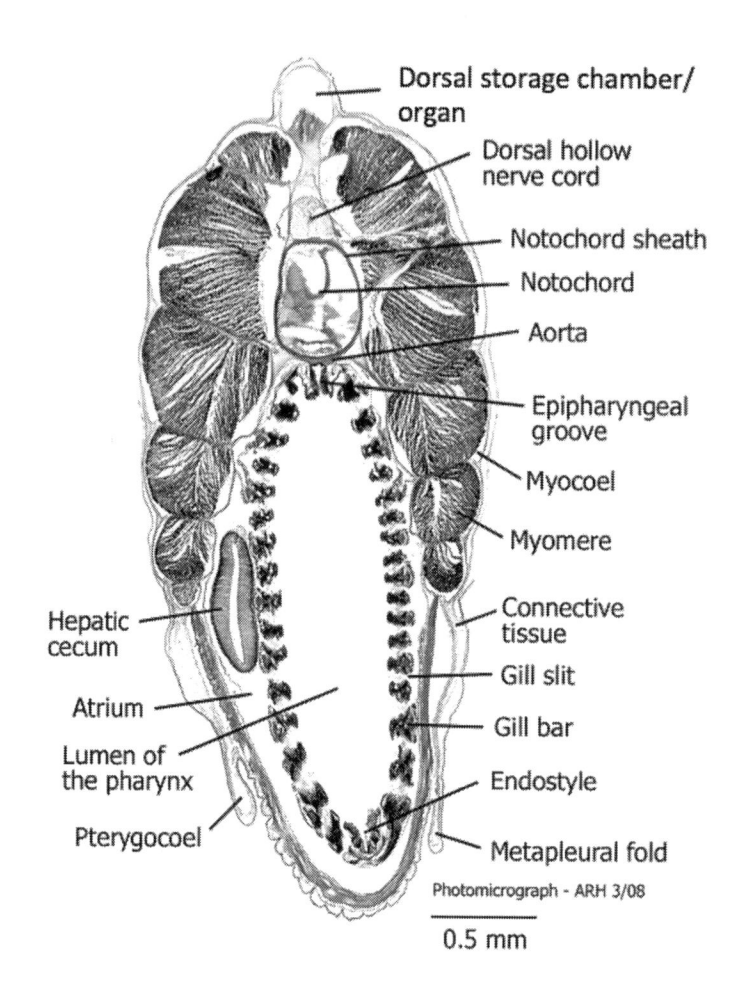

Figure 11.4. Cross-section view through the pharyngeal region of a cephalochordate. (Image: ARH)

Group Questions

1) Develop a hypothesis that explains ecological advantages gained by seastars when they lost their bilateral symmetry as adults in favor of radial symmetry.
2) Develop a hypothesis that explains evolutionary advantages gained by seastars when they lost their bilateral symmetry as adults in favor of radial symmetry (Hint: think reproductive advantage).
3) List and describe the functions of each of the five defining anatomical characteristics of Phylum Chordata.

Phylum Echinodermata – Glossary

Aboral surface – surface of the body opposite the surface that bears the mouth

Ambulacral groove – depression that runs along the aboral surface of arms of seastars, houses the radial nerve and radial and lateral canals of the water vascular system

Ambulacral ridge – raised ambulacral structure as viewed from within a seastar arm, houses the radial canal of the water vascular system and radial nerve

Ampulla – the part of a tube foot that is inside the body, it fills with water when the podium contracts and contraction of muscles surrounding the ampulla extends the podium

Aristotle's Lantern – complex chewing mouthpart of some echinoids

Bivium – the two arms of a seastar associated with the madreporite

Cardiac stomach (seastar) – portion of the stomach that can be extended out through the mouth and is the portion of the digestive tract where food is stored temporarily before being moved to the pyloric stomach

Cardiac stomach retractor muscles – these muscles pull the cardiac stomach back into the body through the mouth

Central disc – structure located in the center of crinoids, asteroids and ophiuroids, bears the mouth and anus and gives rise to the arms

Larva – developmental stage between hatching and metamorphosis into the juvenile stage

Lateral canals – short canals of the water vascular system that carry water from the radial canals to the tube feet

Madreporite – a sieve plate through which water is pulled into the water vascular system

Mutable connective tissue – connective tissue that can change consistency between nearly liquid and rigid as needed

Pentaradial symmetry – body plan where the body produces five sets of structures that radiate from a central axis

Podia – portion of the tube feet that extend beyond the body wall

Pyloric cecum – digestive gland

Pyloric stomach (seastar) – portion of the stomach that moves food from the cardiac stomach to the digestive cecae for digestion and absorption

Radial canal – canal of the water vascular system that carries water from the ring canal supplies water to the lateral canals

Ring canal – structure of the water vascular system that carries water from the stone canal to the radial canals

Stellate – star-shaped

Stone canal – structure of the water vascular system that carries water from the madreporite to the ring canal

Suspension feeding – feeding strategy where floating particles are captured and ingested

Test – the rigid endoskeleton of some echinoderms

Trivium - three arms of a seastar that are not associated with the location of the madreporite

Tube feet – portion of the water vascular system used for locomotion, feeding and sensing the environment, may include external podia and internal ampulla or just podia, varies by taxon

Water vascular system – unique hydraulic system of channels, canals and podia found only in echinoderms

Phylum Chordata – Glossary

Aorta – also called the dorsal vessel

Atriopore – water that passes through the pharynx enters the atrium and leaves the body via this excurrent opening in lancelets

Atrium – fluid-filled space surrounding the pharynx of cephalochordates

Dorsal hollow nerve cord – tube of ectodermal origin that forms by infolding and gives rise to the brain and nerve cord, differs from nerve cords of other invertebrate phyla that are ventral and solid

Endostyle/thyroid gland – glandular organ located along the ventral wall of the pharynx, secretes the hormone thyroxin and in invertebrate chordates also secretes sheets of mucus used for suspension feeding, the endostyle is homologous to the vertebrate thyroid gland

Iliocolon – portion of the digestive tract of cephalochordates where most extracellular digestion takes place

Infaunal – living partially or completely buried in the substrate

Notochord – stiff but flexible dorsal rod that provides structural support and is mesodermal in origin, it has a sheath that houses many disc-shaped elements

Oral cirri – tentacle-like structures located at the opening of the buccal cavity in cephalochordates, these are sensory and keep large particles from entering the pharynx

Pharyngeal gill slit/opening – openings in the pharynx used for respiration and feeding

Post-anal tail – muscular tail the extends beyond the anus

Rostrum – anterior portion of the cephalochordate head located above the oral hood

Velum – transverse sheet of tissue bearing an opening that joins the buccal cavity and the pharynx and regulates what enters the pharynx

Wheel organ – organ located in the buccal cavity, has ciliated stubby finger-like projections and its cilia together with cilia lining the pharynx pulls water into the pharynx

Chapter 12: Animalia – Vertebrates

Subphylum Vertebrata includes around 70,000 species that have a cranium that protects the brain and a supporting notochord or vertebral column. Vertebrates have more copies of *Hox* genes than other animals, they are more anatomically complex and are, in most cases, larger than other animals.

Vertebrates are the animals most people are familiar with. If you were to conduct an experiment and asked 100 randomly selected people to name 10 animals each, the vast majority of those animals would be vertebrates and most of those would be mammals.

Vertebrates are divided into eight classes. These classes are listed in **Table 12.1**.

Table 12.1. Classes in Subphylum Vertebrata (Craniata).

Class name	Common name or representative	Habitat and ecology	Image
Myxini	Hagfish	Deep sea benthic, scavengers and predators	
Petromyzontida	Lamprey	Marine and freshwater, ectoparasites and predators	
Chondrichthyes	Sharks, skates and rays	Marine and freshwater, mainly predators and scavengers	
Actinopterygii	Ray-finned fishes	Marine and freshwater, mainly herbivores and predators	
Sarcopterygii	Coelocanths and lungfishes	Marine and freshwater, predators	

Amphibia	Amphibians	Freshwater and terrestrial herbivores and predators	
Reptilia	Reptiles, including birds	Marine, freshwater and terrestrial, herbivores and predators	
Mammalia	Mammals	Marine, freshwater and terrestrial, mainly herbivores and predators	

Class Myxini (myxin = a slime fish)

There are 70 species of hagfishes. Hagfishes are indeed slimy as the class name suggests; it is one of their defining characteristics. When they are perturbed, hagfishes release a secretion into the water that immediately swells into a massive amount of slime that they can use to protect themselves. Hagfishes have a cartilaginous **cranium** and **notochord** but they lack paired appendages and jaws, and there is still some debate about whether the notochord is divided into vertebrae or not. Hagfishes are chemosensory scavengers that have 3 or 4 pairs of tentacle-like **barbels** around the mouth. They feed by using a rasping tongue to remove tissue from dead animals they find on the sea floor.

Class Petromyzontida (petro = rock, myzo = suck, ida = like)

There are 40 species of lamprey. Many species of lamprey spend their adult lives at sea but enter freshwater to spawn. Their larval and juvenile life stages of these species live in rivers and streams. Others species live their entire lives in freshwater. All lamprey are jawless, eel-shaped animals with a round mouth they use to attach themselves to larger animals where they live as ectoparasites. Juveniles living in rivers sometimes attach themselves to rocks, thus the source of the class name – rock suckers. Like hagfish, lamprey lack paired appendages, are jawless, and have a cartilaginous cranium and vertebral column.

Class Chondrichthyes (chondro = cartilage, ichthy = fish)

There are 1000 species in this class, including sharks, skates, rays and relations. These fishes have a cartilaginous cranium and skeleton that in some species can be hardened with calcium carbonate, but is not true bone. They differ from hagfish and lamprey because have paired fins and jaws with teeth. Their teeth are not in sockets but are held in place by soft tissue. Most members of this class are highly specialized predators that find their prey using a combination of chemosensory organs, a **lateral line** system that senses vibrations in the water, image-forming eyes, and **Ampullae of Lorenzini** that are organs in the **rostrum** that are sensitive to electrical signals produced by living things.

Class Actinopterygii (actin = ray, pter = wing)

These are the ray-finned fishes. This is the most divers class of vertebrates with over 30,000 species, and they are found in virtually every aquatic habitat. These fishes differ from hagfish, lamprey and sharks because they have a true bone **cranium** and skeleton and a **swim bladder** that is a lung-precursor. Fins of these fishes have stiff rays but fleshy tissue does not extend into them.

Class Sarcopterygii (sarco = flesh, pter = wing)

These are **lobe-finned** fishes and include Subclass Actinistia (coelacanths) and Subclass Dipnoi (lungfishes). Like ray-finned fishes, lobe-finned fishes have bony skeletons but unlike ray-finned fishes, fleshy tissue and muscles extend into their fins. Coelacanths are living fossils because they were known only from fossils until a living specimen was identified in 1938. Scientists then discovered that local fishermen knew these large fish (two meters long), but they usually threw them back when they were caught because they are not good to eat. Living coelacanths are known to exists only off the eastern coast of Africa and in waters of Indonesia. Lungfishes on the other hand live in freshwater. There are six known living species of lungfishes. They live in tropical South America, tropical Africa and eastern Australia. Lungfishes have bony skeletons and craniums, and their swim bladders are highly vascularized and are used for extensively for gas exchange. Some lungfishes will even drown if they are prevented from reaching the surface. Phylogenetic analysis revealed that lungfishes are the sister taxon to **tetrapod** vertebrates.

Class Amphibia (amphi = both sides or double, bio = life)

There are 6,800 species of amphibians, including frogs, salamanders and ceacilians. Amphibians are the basal taxon of **tetrapods** – animals with four legs. There is an amazingly good fossil record showing the evolutionary link between lobe finned fishes and amphibians. Amphibians have a bony skeleton and pectoral and pelvic girdles with legs that support the body. Their skin is moist, permeable and **glandular.** Amphibians will quickly desiccate if they cannot stay moist. Amphibian larvae are usually herbivorous but all adults are carnivores that will eat just about anything.

Class Reptilia (reptil = creep or crawl)

Over 9,500 species of reptiles exist today. Living species include lizards, snakes, turtles, crocodiles, birds and their relations. This class also includes many extinct species (none of which are covered in this lab manual), including dinosaurs, ancient marine reptiles and pterosaurs. Interestingly, crocodiles and birds are surviving members of the evolutionary line including dinosaurs. So, in a way, there are still dinosaurs among us!

Reptiles differ from amphibians in multiple ways. Reptiles have skeletons that are heavier and more extensive than amphibians, most of them have **homodont teeth** in sockets in the jaw, their bodies are covered by **scales** or **feathers**, and they produce **amniotic eggs**. Historically, birds were assigned to their own class (Class Aves) but phylogenetic analysis shows that birds are derived reptiles, so they are now included in Class Reptilia.

Class Mammalia (mamill = teat)

There are over 5,000 species of mammals. Mammals differ from reptiles in that they have **hair**, **mammary glands**, glands in the skin that produce oil, sweat, and other secretions, a **diaphragm** used during breathing, three middle ear bones, and **heterodont dentition** with teeth in sockets.

Subphylum Vertebrata

Though there is a great deal of morphological diversity within Subphylum Vertebrata, the internal anatomy of vertebrates is amazingly similar. That is, the major internal organs are pretty much in the same places and have similar anatomical orientations to each other regardless of class. You will discover this during this lab exercise. For example, the **lungs** and **kidneys** are usually dorsal, the **heart** is ventral and anterior, the **liver** is also anterior, the **pancreas** is located in a loop of the gut and is attached to the **pyloric sphincter** that is located at the anterior end of the small intestine.

The main goal of this laboratory exercise is to give you the chance to study the external and internal anatomy of representative vertebrates. Representatives of classes Chondrichthyes, Actinopterygii, Amphibia, Reptilia and Mammalia are included in this exercise. After you study internal, external and in some cases, skeletal anatomy of your specimen, compare what you saw with the external and internal anatomy other representatives in this chapter.

During this lab period you will study one or more representative vertebrate in detail and then compare what you saw with other material available in this chapter. You should WRITE a list of similarities and differences between the animal you studied and the others in this chapter.

Task #1: Class Chondrichthyes – spiny dogfish shark *Squalus*

1) External anatomy
   a. Sharks are very oily and smelly, so make sure that you roll up long sleeves, put up long hair and wear gloves before you start this dissection.
   b. Obtain a large dissection tray and a preserved shark. Observe, describe and DRAW the external anatomy of your shark. Refer to **Fig. 12.1** to help you identify what you see.
2) Internal anatomy – shark dissection
   a. Use a pair of heavy scissors to cut through the angle of the jaw on both sides and open the oral cavity. Look for internal gill openings and the opening of the spiracle. WRITE down what you see.
   b. Lay the dogfish on its back and, using a pair of heavy scissors, make a longitudinal cut starting at the anus and continuing anteriorly until you reach a point between the pectoral fins.
   c. Pin down or cut away the body wall to expose the organs of the body cavity.
   d. Rinse out the body cavity of the shark at a sink and examine the orientation of the

visible internal organs before moving any of them.
e. Next, move aside the large lobes of the liver to expose deeper organs as shown in **Fig. 12.3** and examine the organs formerly covered by the liver (*do not remove any organs*).
f. Next, move aside the organs of the digestive system so you can see the reproductive organs and kidneys along the dorsal body wall on both sides of the body cavity. Again, do not remove any organs, just gently pull them to one side to reveal these deeper organs.
g. Make a circular cut on the underside of the head just anterior of the pectoral girdle to expose the heart, ventral aorta and gills as shown in **Fig. 12.3**.
h. DRAW your dissected specimen and use **Fig. 12.3** to help you identify what you see.

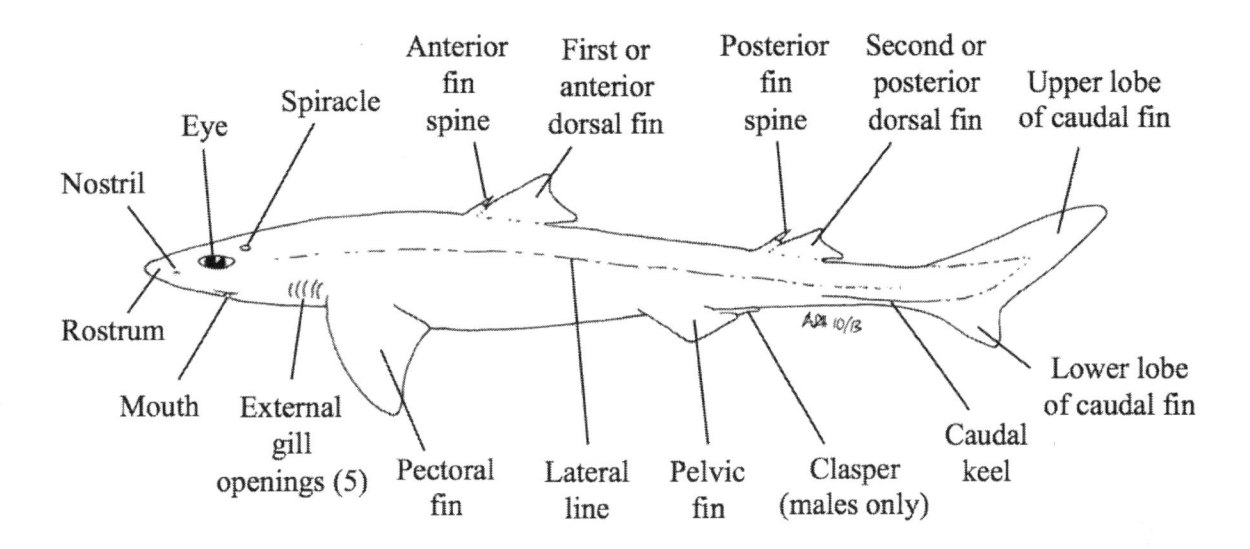

Figure 12.1. External anatomy of the dogfish. Note: Many sharks have an anal fin posterior to the pelvic fins and the cloaca, but dogfishes do not. (Image: ARH)

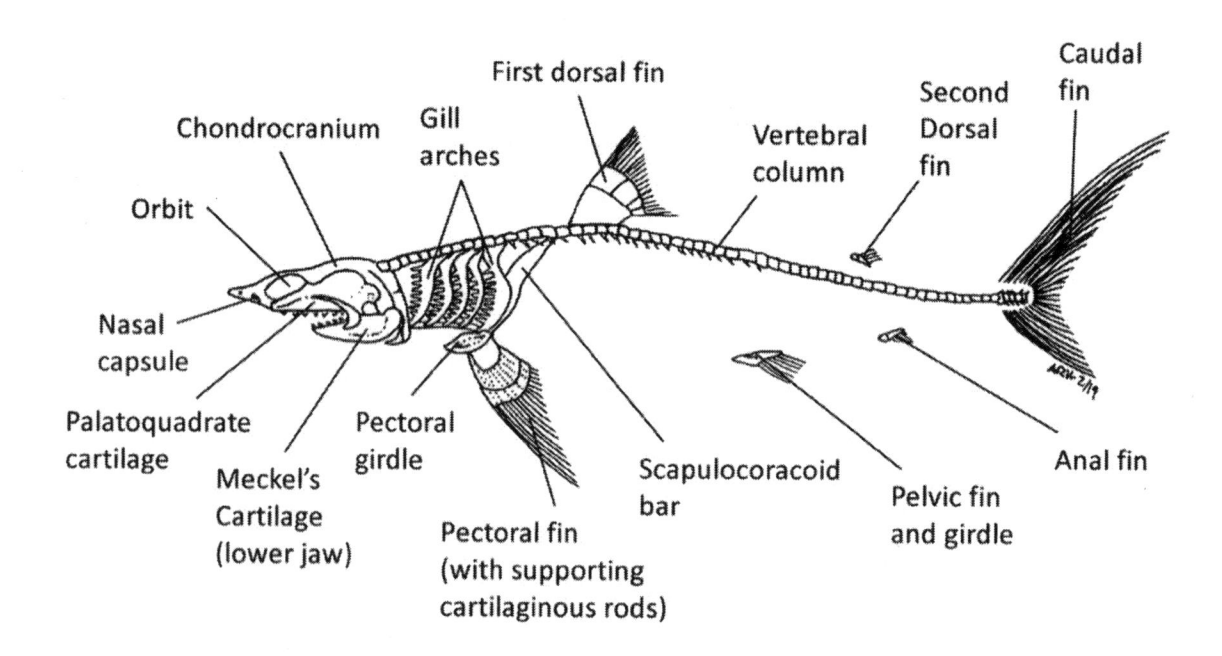

Figure 12.2. Cartilaginous skeleton of a white shark. (Image modified by ARH from fair-use image, https://www.fotolibra.com/gallery/1201414/shark-skeleton/ Miles Kelly Collection)

3) Shark skeletal anatomy
    a. Study **Figure 12.2** and pay particular attention to the anatomy of the head, jaw, paired fins and backbone.

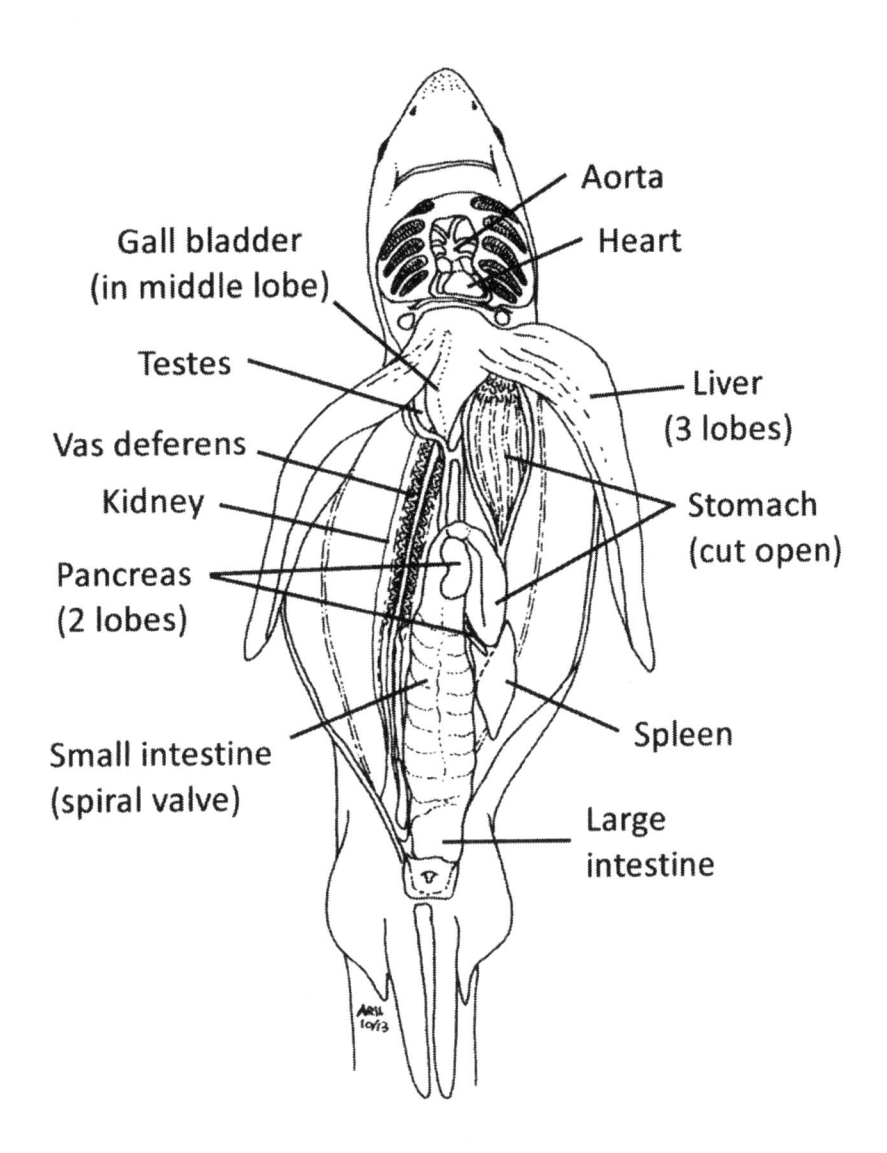

Figure 12.3. Internal anatomy of the dogfish shark. Note: The heart has one atrium and one ventricle. (Image: ARH)

Task #2: **Class Actinopterygii – a ray-finned fish, the perch, *Perca***

1) Bony fish external anatomy. Obtain a preserved perch. Observe, describe and DRAW your specimen. Use **Fig. 12.4** to help you identify what you see.
2) Bony fish skeletal anatomy. Either study a mounted skeleton of a bony fish or **Figure 12.5** to help you identify structures of the bony fish skeleton. Do not take time to draw the skeleton. Pay particular attention to the anatomy of the head, jaw, paired fins and backbone.

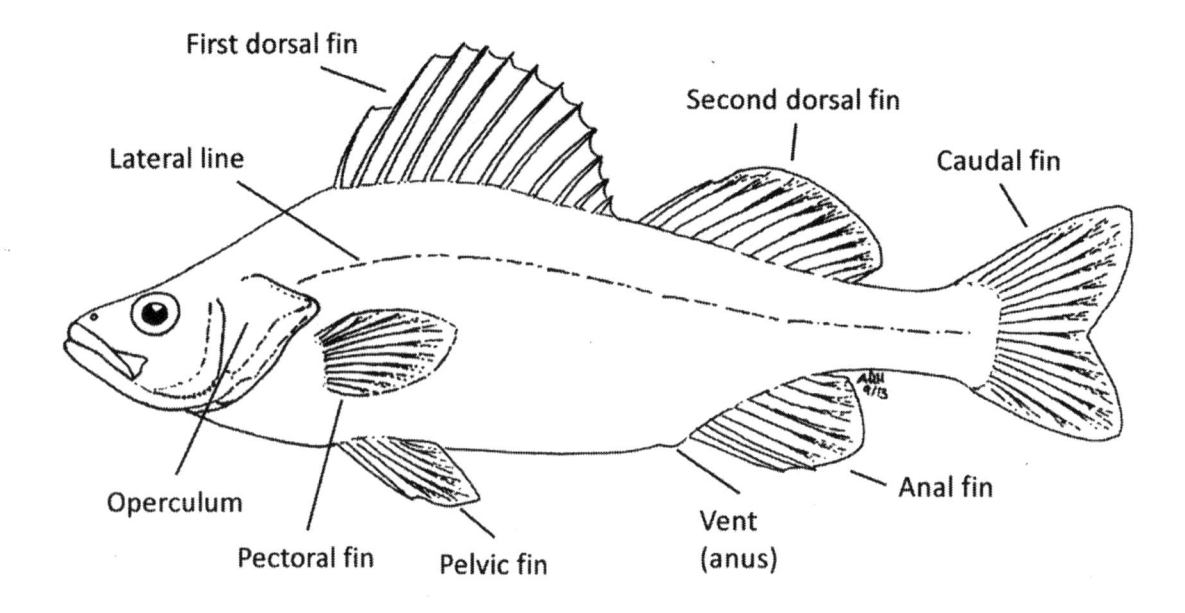

Figure 12.4. External anatomy of the perch. (Image: ARH)

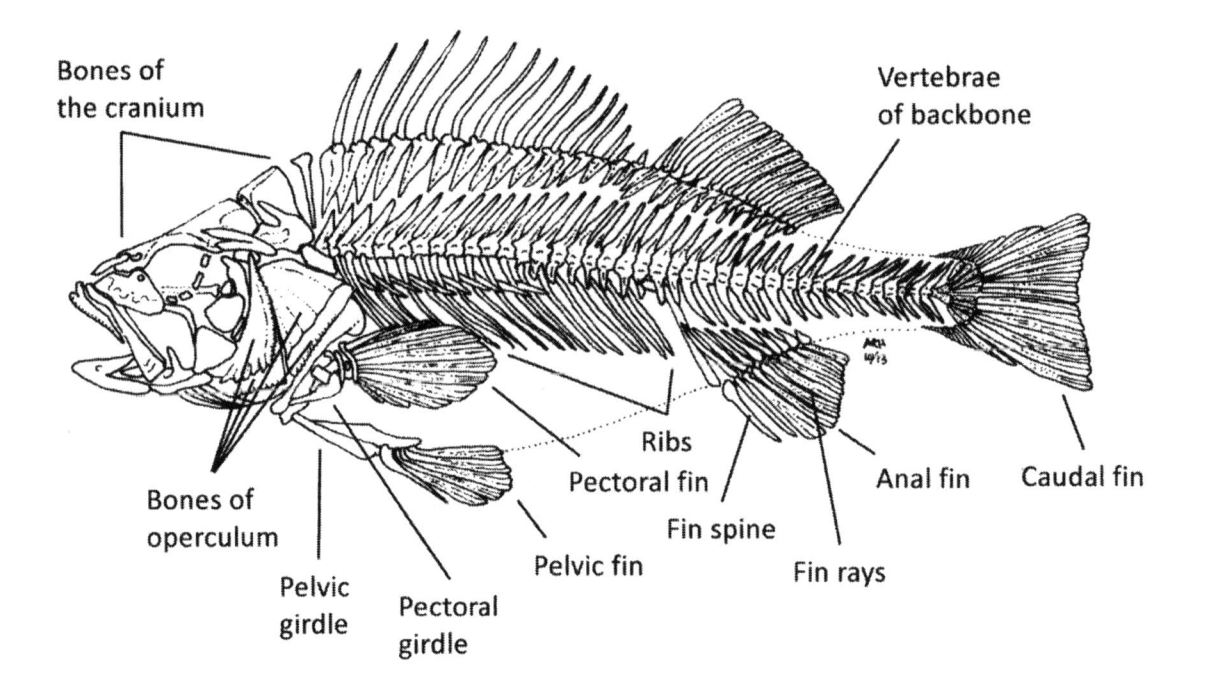

Figure 12.5. Skeletal anatomy of the perch. Note that pectoral, pelvic, anal and second dorsal fins have a fin spine and many fin rays, and that this fish has pectoral and pelvic girdles to support those paired fins. (Image by Alan Holyoak based on the image by Albert Günther, Figure 23 Skeleton of Perch in *An Introduction to the Study of Fishes*, 1880, public domain).

3) Bony fish internal anatomy. Dissection instructions.
    a. Lay your specimen on its right side and use sharp-tip scissors to make an incision starting at the vent (anus), cut dorsally. Continue the incision anteriorly, then

back around to the anus producing a large window, exposing the organs of the body cavity as shown in **Fig. 12.6**. Examine the organs in the body cavity before removing any organs.

b. After looking at the organs in place, remove the liver and gall bladder in order to expose the organs underneath. Refer to **Fig. 12.6** to see what your fish should look like at this point.

c. Cut away the operculum (bony covering of the gills) and the tissue covering the heart and ventral aorta. See **Fig. 12.6**.

d. Describe and DRAW your dissected specimen. Refer to **Fig. 12.6** to help you identify what you see.

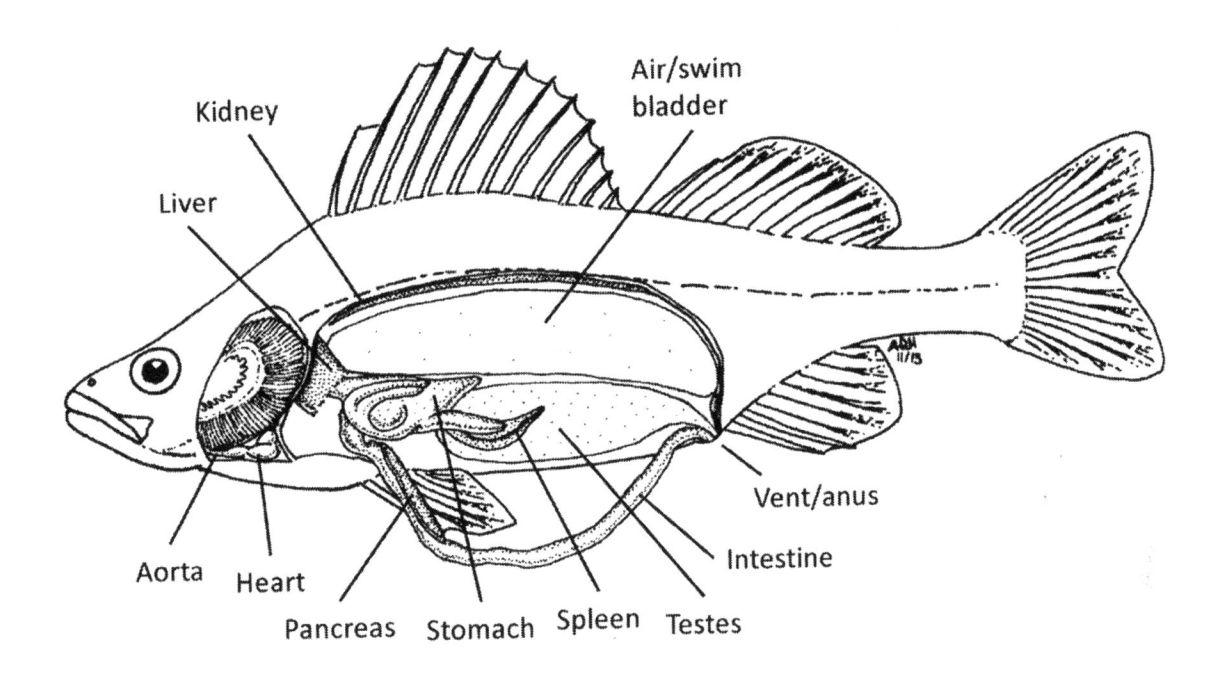

Figure 12.6. Internal anatomy of the perch. Note that the liver and gall bladder are cut away so they no longer obscure other major organs. (Image: ARH)

Task #3: Class Amphibia – the mudpuppy salamander, *Necturus*

2) Salamander external anatomy. Obtain a preserved specimen, then observe, describe and DRAW it. Use **Fig. 12.7** to help you identify what you see. Also, look between the external gills to observe the external gill openings.

3) Salamander skeletal anatomy. Study a mounted skeleton of a salamander if available. If one is not available, study **Fig. 12.8**. Do not take time to draw the skeleton. Pay particular attention to the cranium, jaw, appendages and backbone. Write down your observations.

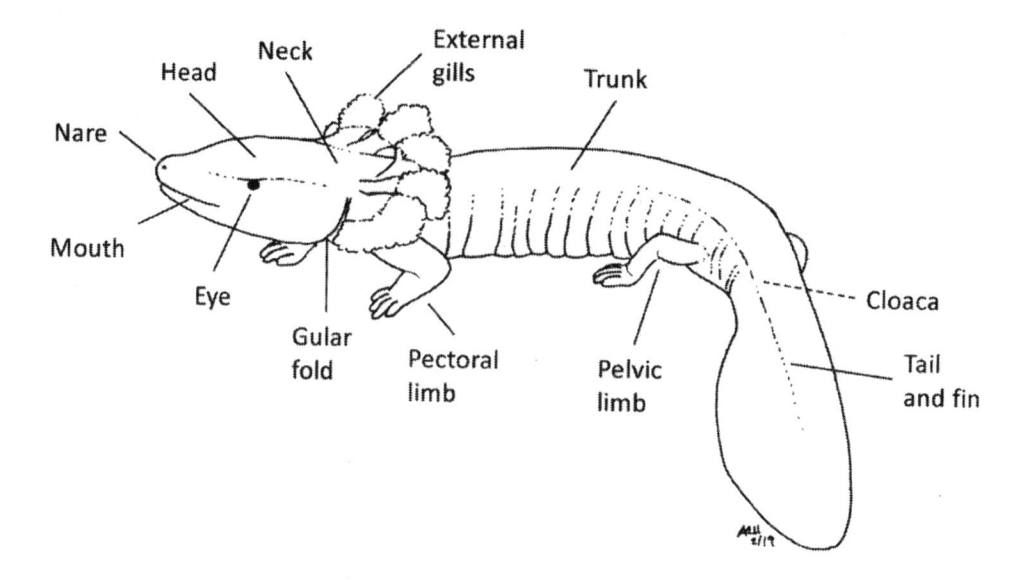

Figure 12.7. External anatomy of *Necturus maculosus*, the mudpuppy. **Note**: The cloaca is indicated with a dashed line because it is located ventrally and just posterior to the pelvic limbs. (Image: Modified by ARH from http://www.uvm.edu/~jbartlet/amphibian/mudpuppy.jpe)

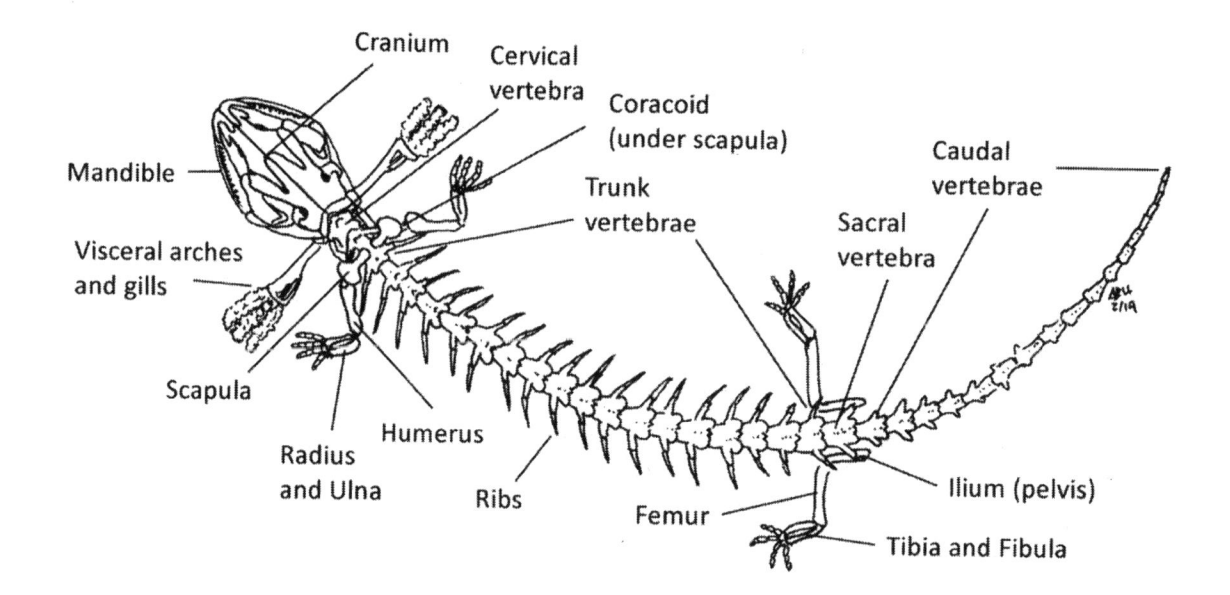

Figure 12.8. Dorsal view of the skeletal anatomy of the salamander *Necturus*, the mudpuppy. **Note**: There is typically one cervical and one sacral vertebra in adult amphibians. Carpals and metacarpals of the forelimb are not indicated, neither are the tarsals and metatarsals of the hind limbs. (Image: ARH)

4) Internal anatomy of the salamander. Dissection instructions.
   a. Oral cavity. Use a pair of heavy scissors to cut through the angle of the jaw on both sides of the head. Continue these cuts until you can open the entire oral

cavity. Look for the internal gill openings. WRITE observations about what you see in the oral cavity.

b. Open the main body cavity. Use sharp-tip scissors to make a shallow, longitudinal incision off-center of the ventral midline to open the main body cavity, also called the **pleuroperitoneal cavity**. Start the incision at the cloacal opening and continue cutting until you reach a point just posterior to the gular (neck) fold. Pull the body wall open and rinse the body cavity. Use paper towels to remove any residual fluid.

c. Make lateral cuts in the body wall as needed and pin the body wall to the bottom of a wax-bottom dissection tray or cut away the body wall so it will not obstruct the view of the internal organs.

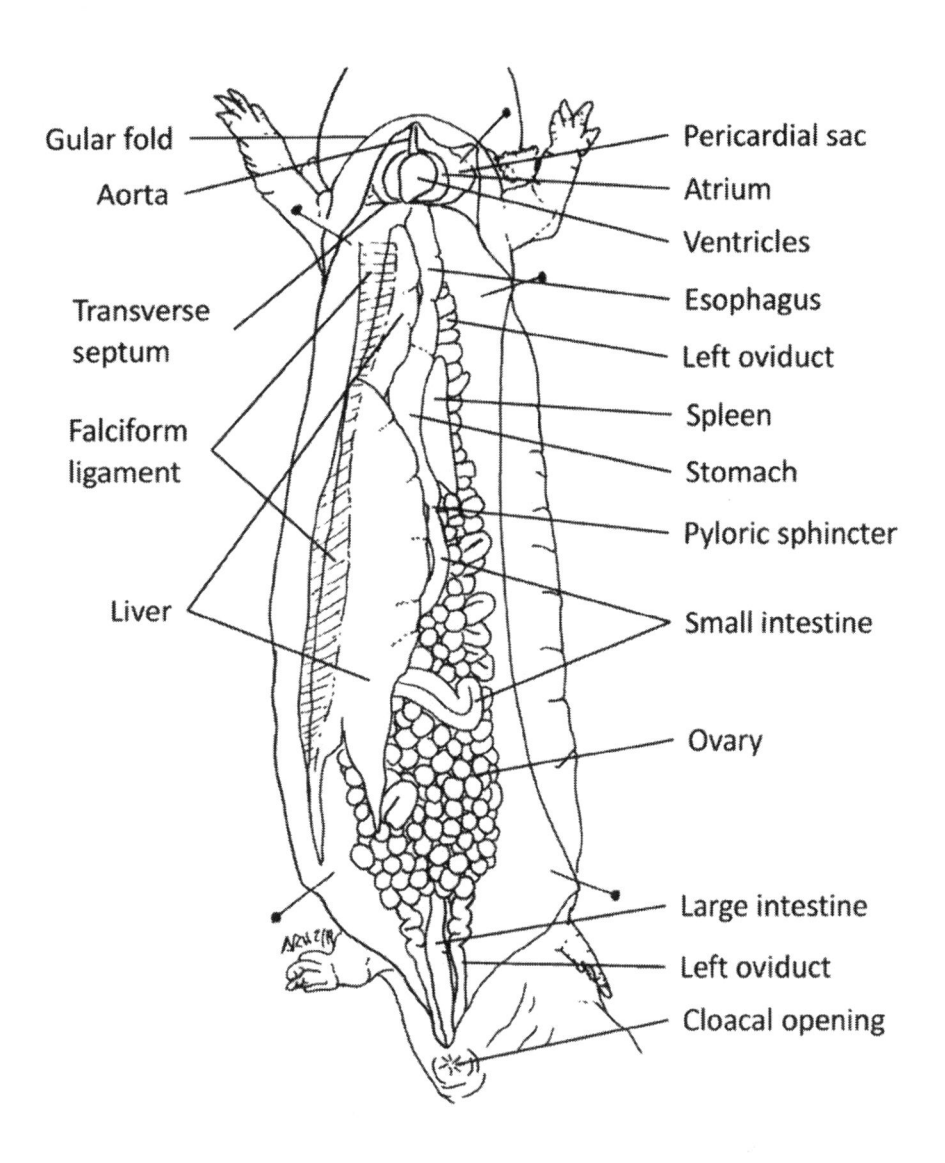

Figure 12.9. Organs of the pericardial and pleuroperitoneal cavities of the salamander *Necturus*, prior to any manipulation of any organs. (Image: ARH)

d. With the pleuroperitoneal cavity open, examine the superficial organs. Look first for a large mottled greenish-gray, leaf-shaped organ that wraps around organs along the animal's right-hand side. This is the **liver**. It is attached to the body wall by a long ligament called the **falciform ligament**. You should be able to see portions of the **esophagus**, **stomach**, **spleen**, **small intestine**, **oviduct** and **ovary** along the left-hand side of the body cavity. See **Fig. 12.9** to help you identify what you see.

e. Examine the deeper organs in the pleuroperitoneal cavity. Gently move (but do not remove) the liver to the salamander's right-hand side and pin it there. The liver can be quite delicate and will tear or break if you are not careful. You should now be able to see organs that were previously obscured. With the liver moved out of the way, carefully remove the right-hand set of reproductive organs: the ovary and oviduct. Use **Fig. 12.10** to help you see what your specimen should look like.

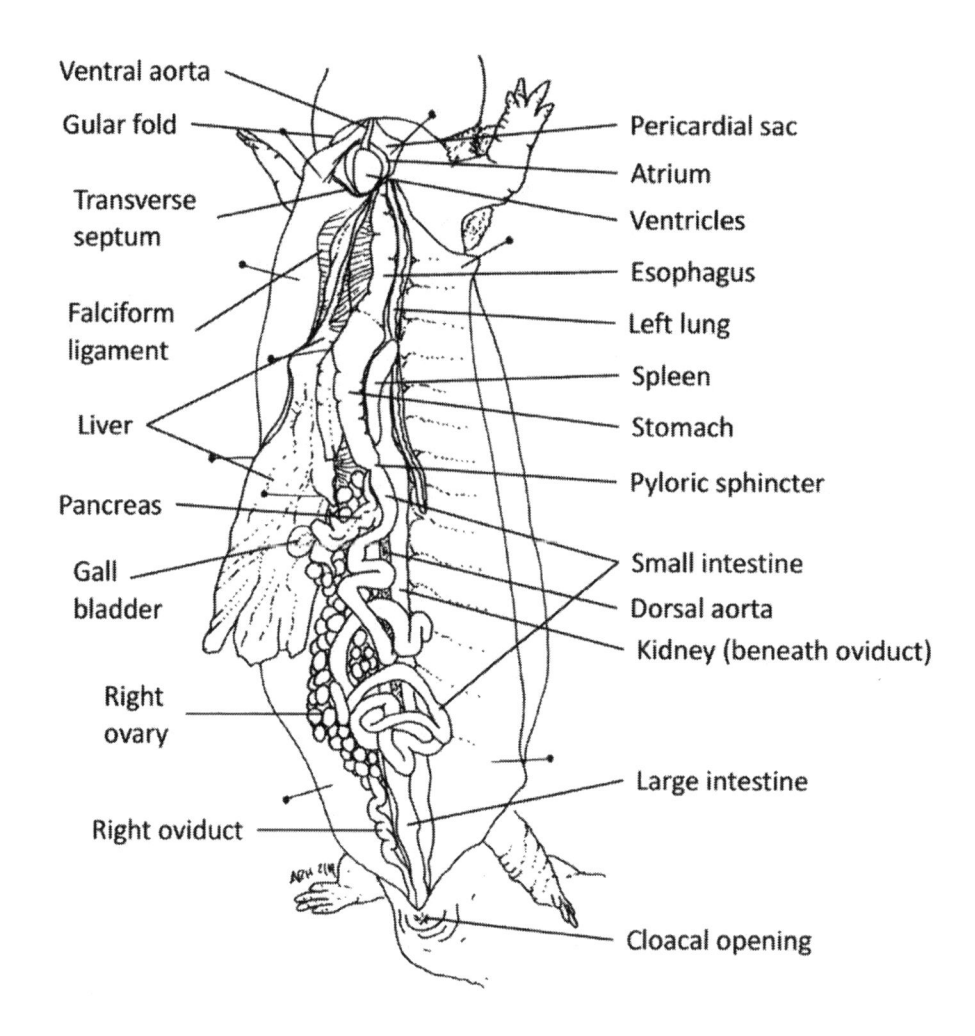

Figure 12.10. Anatomy of the pleuroperitoneal cavity of a female *Necturus*, the mudpuppy, with the liver displaced to the right-hand side of the specimen and the left-hand set of reproductive organs removed. (Image: ARH)

f. Open the pericardial cavity located anterior to the pleuroperitoneal cavity. It houses the heart consisting of left and right atria, a single ventricle and ventral aorta. The pericardial cavity is located between the pectoral girdle and the gular fold. It is isolated from the pleuroperitoneal cavity by a layer of tissue called the transverse septum. If needed, make additional cuts in the body wall to fully expose the pericardial cavity. Refer to **Fig. 12.10** to see what your specimen should look like.

g. DRAW your dissected specimen and use **Figs. 12.9-10** to help you identify what you see.

Task #4: Class Reptilia – a lizard, *Anolis*

1) Lizard external anatomy.
   a. Obtain a preserved specimen and observe, describe and DRAW what you see. Use **Fig. 12.11** to help you identify structures on your specimen.
   b. Like geckos, *Anolis* lizards can crawl up seemingly impossible surfaces. Use a magnifying glass or dissection scope to examine the toe pads. WRITE your observations about the toe pads and hypothesize about how they allow these lizards to climb the way they do.

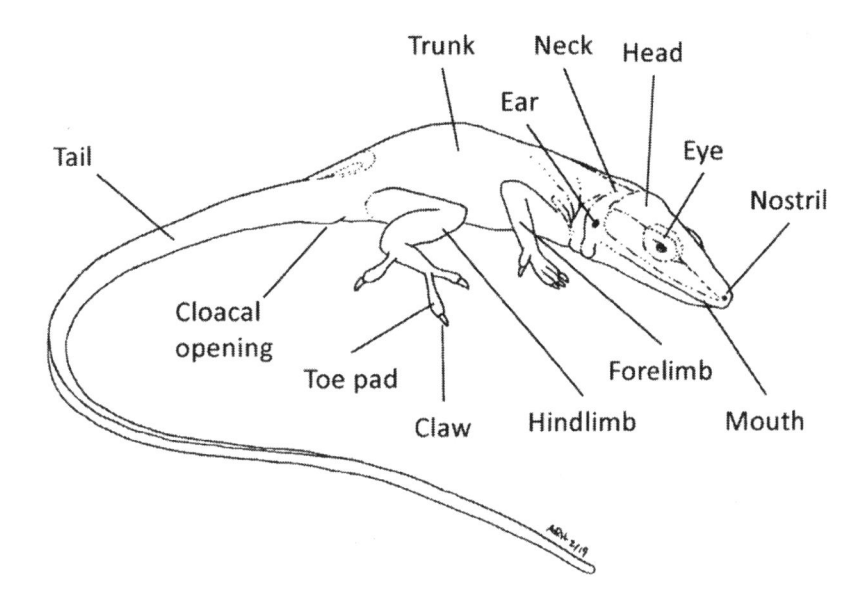

Figure 12.11. External anatomy of the *Anolis* lizard. (Image: ARH)

2) Lizard skeletal anatomy.
   a. Study a mounted skeleton of a lizard if available. If one is not available, study **Fig. 12.12**. Do not take time to draw the skeleton. Pay particular attention to the head, jaw, backbone and paired appendages. Describe what you see.

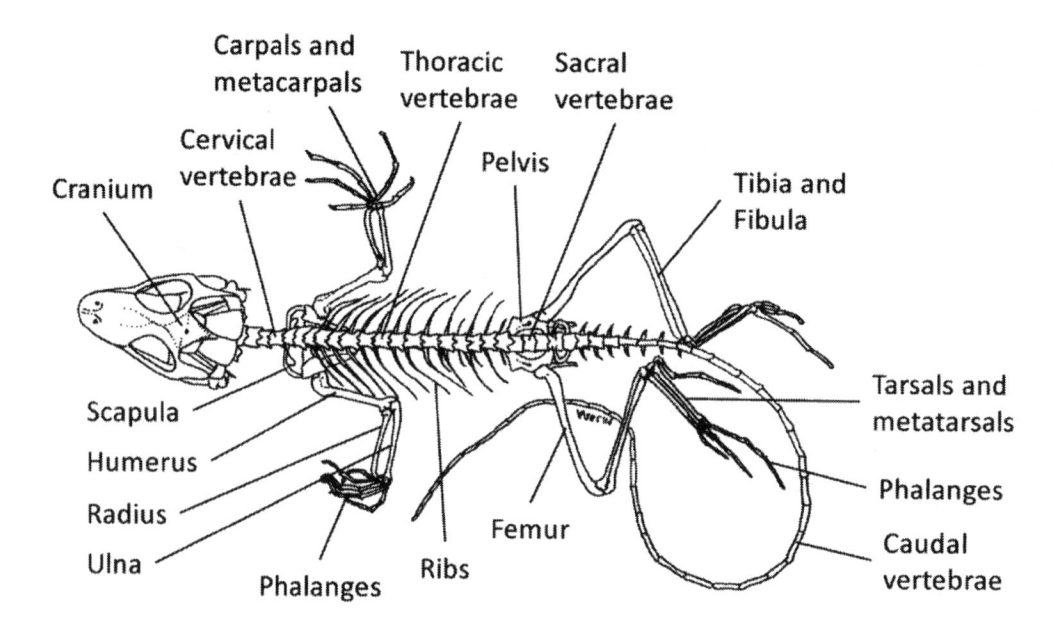

Figure 12.12. Skeletal anatomy, dorsal view, of the *Anolis* lizard. (Image: Modified by ARH from http://www.anoleannals.org/2011/11/18/anolis-now-in-3d/)

3) Lizard internal anatomy.
   a. Pin your specimen ventral side up, to a wax-bottom dissection tray.
   b. Use sharp-tip scissors to make a longitudinal incision starting at the cloacal opening and continuing anteriorly beyond the pectoral girdle.
   c. Cut away the body wall, forelimbs and as much of the pectoral girdle as is necessary to expose the heart.
   d. Examine superficial organs of the main body cavity without moving any organs. The left-hand image of **Fig. 12.13** will help you identify what you see before any organs are displaced. Most of the body cavity is obscured by large lobes of the liver anteriorly and lobes of the kidney posteriorly.
   e. Remove the liver and kidneys so you can see the deeper organs of the body cavity.
   f. DRAW your dissected specimen and use **Fig. 12.13** to help you identify what you see.

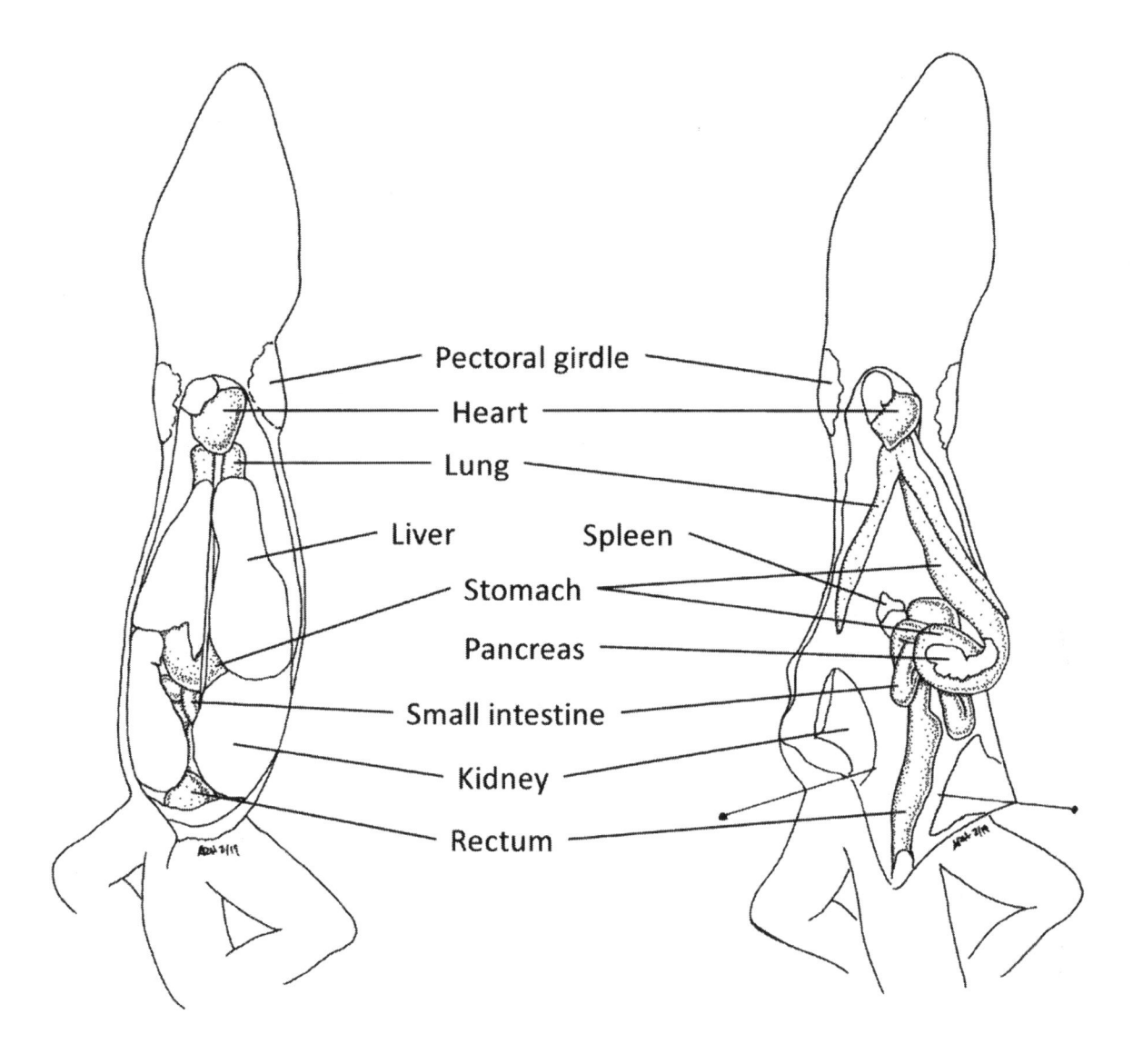

Pectoral girdle
Heart
Lung
Liver Spleen
Stomach
Pancreas
Small intestine
Kidney
Rectum

Figure 12.13. Ventral view of the internal anatomy of the lizard *Anolis*. Forelimbs were removed along with part of the pectoral girdle to expose the heart. **Left:** Major organs of the body cavity in place. **Right:** Liver and kidneys removed to expose deeper organs. (Images: ARH)

Task #5: Class Reptilia – bird, the starling, *Sternus vulgaris*

1) Bird skeletal anatomy. Study a mounted skeleton of a bird if available. If a mounted skeleton is not available study **Fig. 12.14**. Do not take time to draw the skeleton. Pay particular attention to the skull, jaw, backbone and paired appendages. Write down your observations about the bird skeleton. Note adaptations of the skeleton for flight.
2) Bird internal anatomy. Dissection directions.
   a. Opening the body cavity of a bird is different than the procedure for doing so for other vertebrates in this chapter. I'm just warning you ahead of time, this may actually feel a little barbaric, but it's the best and easiest way to open the body cavity.

b. The body cavity is protected by a substantial sternum and ribcage that is not easy to cut through using scissors or anything else. To open the body cavity, make a short lateral incision a few centimeters wide through the loose skin at the base of the sternum. Insert you finger or thumb under the sternum and lift upward. Pull with enough force to make the sternum break away from the ribs and pull the sternum away from the body. You may need to cut away skin on the sides of the sternum to remove it. Remove the sternum and associated tissue completely.
c. Next, expose the trachea and esophagus by cutting away the skin along the ventral surface of the neck. These organs are shallow, so be careful when you remove the skin to avoid damaging them.
d. Observe the organs of the body cavity without displacing any organs. DRAW your specimen and use the left image of **Fig. 12.15** to help you identify what you see. Note that the lobes of the liver obscure the heart and portions of the stomach and other organs.
e. Remove the liver and gall bladder, and displace the organs of the digestive tract toward the left-hand side of the body cavity to expose all parts of the digestive tract as well as deeper organs. DRAW your specimen and refer to the right figure of **Fig. 12.15** to help you identify what you see.
f. Some birds, like pigeons and chickens that eat seeds, have a large thin-walled crop that holds the seeds until they can be moved to the gizzard where they are ground up before moving through to the rest of the digestive tract. Birds that are not seed eaters, like starlings, usually lack a crop.

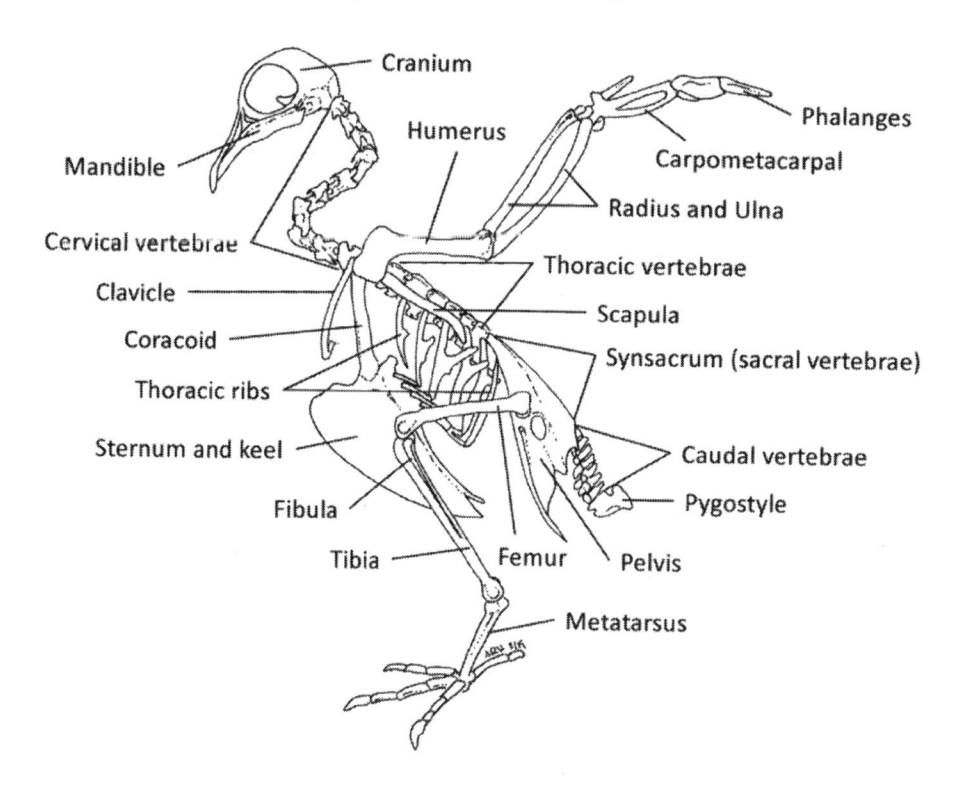

Figure 12.14. Skeletal anatomy of a bird. (Image: ARH)

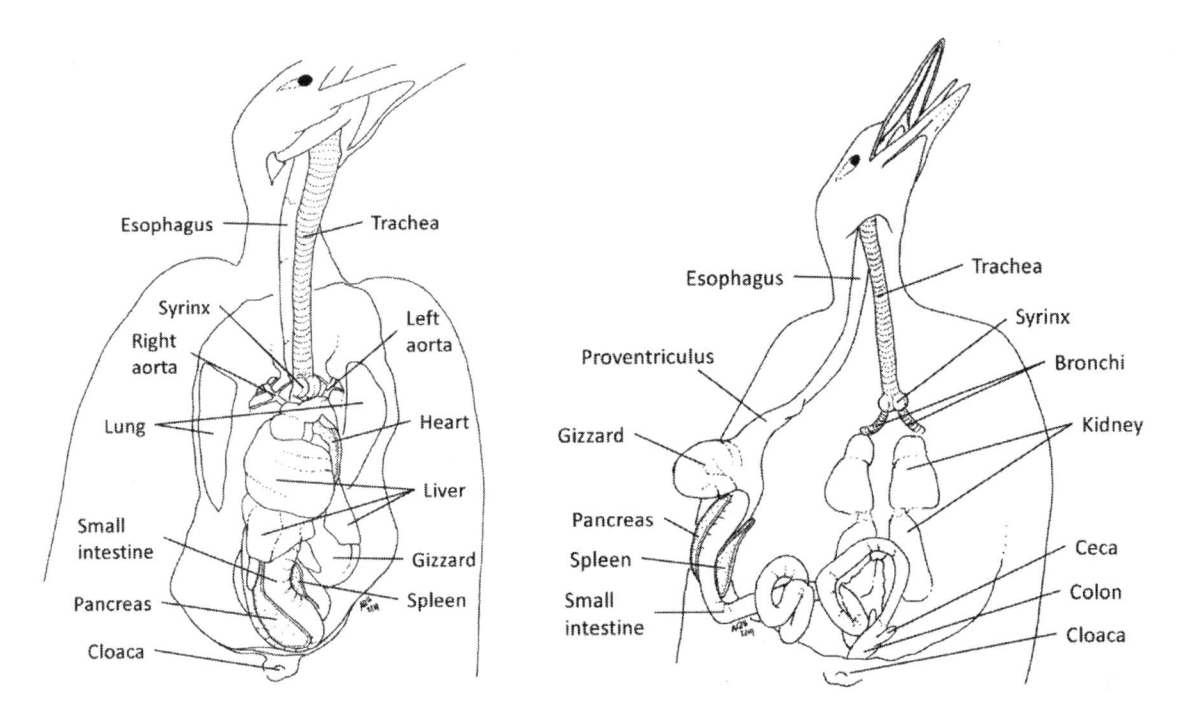

Figure 12.15. Ventral view of the internal anatomy of a starling. **Left:** Body cavity showing internal organs in original orientation. **Right:** Heart and liver removed to expose deeper organs. **Note:** this bird is not a seed-eater and lacks a crop. (Images: ARH)

<u>Task #6:</u> **Class Mammalia, a fetal pig and cat**

1) Cat skeletal anatomy. Study the anatomy of a cat skeleton using a mounted skeleton if one is available. If a mounted skeleton is not available, study **Fig. 12.16**. Do not take time to draw the skeleton. Pay particular attention to the skull, jaw, backbone and paired appendages. Write down observations about how this skeleton facilitates the predatory, hunting lifestyle of a cat (wild, not domesticated).
2) Fetal pig, internal anatomy. Dissection directions.
    a. Obtain a fetal pig and a dissection tray. Tie one end of a length of string and to the foot of one of the forelimbs, pass the string under the dissection tray, spread the legs as far as possible and tie the string to the other foot. Repeat this procedure for the hindlimbs.
    b. Open the main body cavity by making an incision starting at the posterior end of the body cavity and continuing anteriorly, but cut around the opening of the urinary tract, leaving it in place. Continue the incision until it reaches the pectoral girdle.
    c. Cut away the body wall, exposing the organs of the **abdominal cavity**.
    d. Continue your cut anteriorly, exposing the organs of the **pericardial cavity**.
    e. Cut away skin anterior to the pericardial cavity to expose the trachea and esophagus. See **Fig. 12.17** to help you understand where these incisions need to be made and to help you know what your specimen should now look like.
    f. Identify the **diaphragm** that separates the abdominal and thoracic cavities.

g. Identify the major organs of the body cavity. **Fig. 12.17** shows the locations of these organs. DRAW your specimen and identify everything you can see.
h. Remove the liver and displace the intestines to one side to reveal the kidneys which are embedded in the dorsal body wall of the body cavity. They are indicated by the stippled ovals in the posterior end of the body cavity in **Fig. 12.17**. DRAW your specimen again including the kidneys and the entire digestive tract. Use **Fig. 12.17** to help you identity what you see.

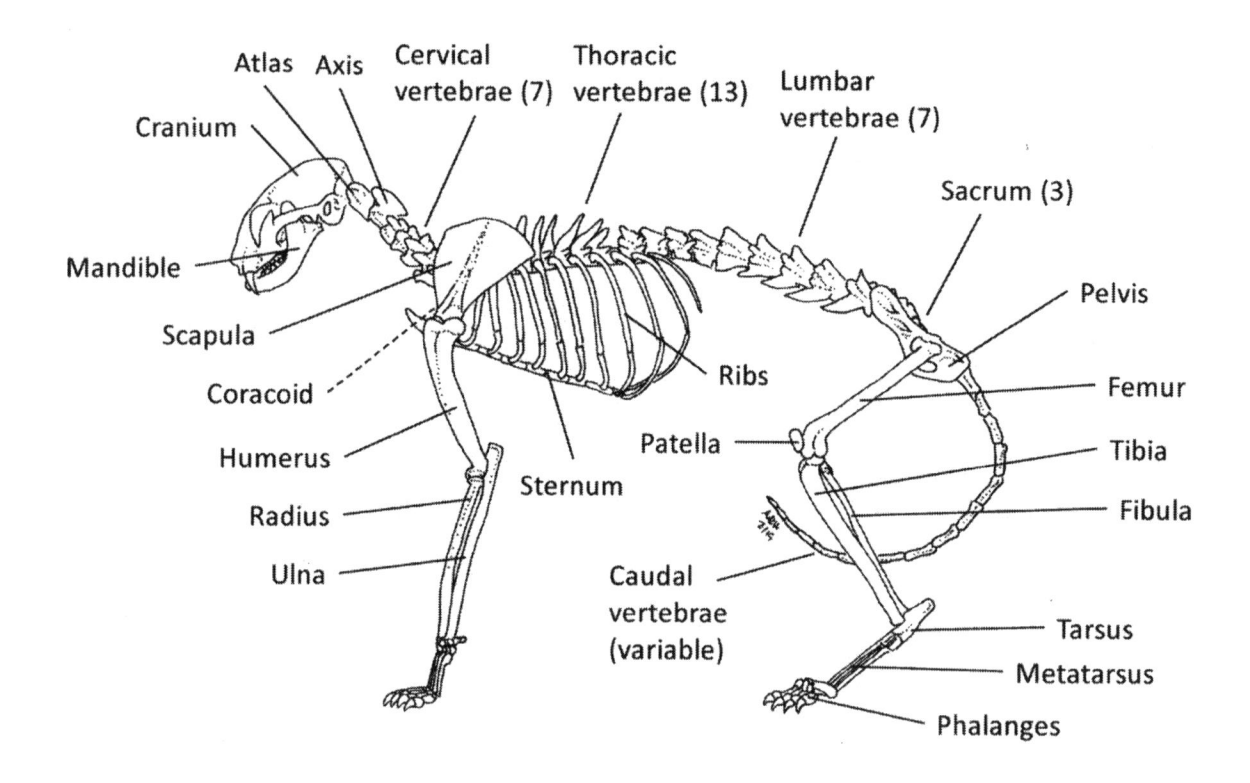

Figure 12.16. Skeletal anatomy of a cat. The coracoid bone is indicated with a dashed line because it is covered by the scapula. (Image: Modified by ARH, from the Museum of Veterinary Anatomy FMVZ USP/Wagner Souza e Silva, Wikimedia Commons)

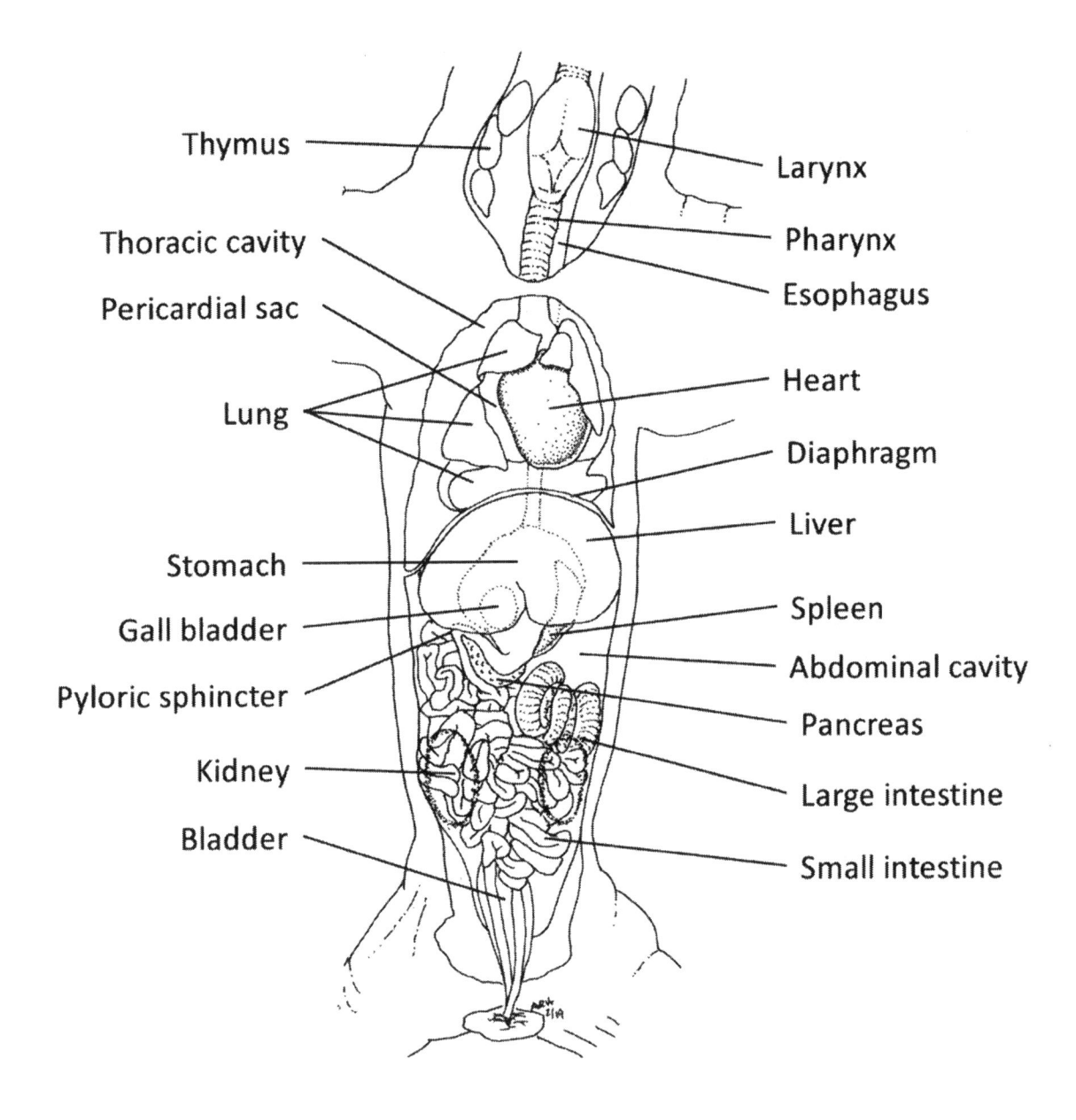

Figure 12.17. Ventral view of the internal anatomy of a fetal pig. (Image: ARH)

Group questions

1) There are over 30,000 species of fishes living in freshwater and marine habitats. 97% of all water on the planet is in the ocean but only 0.03% of water is in rivers and streams (the rest is in ice caps, glaciers and groundwater). However, when we look at the diversity of marine versus freshwater fishes, we find 59% of all fish species live in the ocean and 41% live in freshwater. That's right, 40% live in only 0.03% of the water on Earth! Develop a hypothesis that explains why there are so many species of freshwater fish.
2) Examine all of the skeletons in this chapter, starting with the most ancestral form, the

shark, through the most derived form, the cat. Describe all evolutionary trends between fishes and mammals that you can identify.

3) Explain why the following statement is true: "When you know the internal anatomy of one kind of vertebrate, you generally know the internal anatomy of all vertebrates." Refer to your experience dissecting one vertebrate today, then compare what you observed during that experience with the figures showing internal anatomy of other vertebrates in this chapter.

Subphylum Vertebrata (Clade Craniata) – Glossary

Air/swim bladder – see "swim/air bladder"

Amniotic egg – egg that contains multiple internal membranes including an amniotic sac, a fluid-filled sac that protects the developing embryo

Ampullae of Lorenzini – Electro-sensory organs located on the rostrum and head of sharks, skates, rays and relations

Aorta – largest artery leaving a heart

Atlas – first vertebra of the backbone, supports the head

Atrium – chamber of the heart that receives incoming blood

Axis – second vertebra of the backbone, supports the atlas

Barbel – fleshy, finger-like projections around the mouth of an animal, they are usually tactile and chemosensory organs

Bladder – sac that temporarily stores urine

Bronchi – tubes of the respiratory system that form where the trachea splits

Carpals – bones of the wrist

Carpometacarpal – fused bones of the forelimb of birds, provides more stable wing

Caudal fin – tail fin

Caudal vertebrae – vertebrae of the tail

Ceca – closed pouch that branches off of the digestive tract

Cervical vertebra – vertebrae between the head and the pectoral girdle

Chondrocranium – cranium made of cartilage

Clasper – accessory male reproductive structure found on sharks and their relations

Clavicle – collar bone, connects the shoulder to the sternum, wishbone in birds

Claw – bony structure that protects the tip of fingers or toes

Cloaca – opening through which products of more than one organ system are released from the body, e.g., digestive and reproductive

Colon – terminal structure of the digestive tract (other terms exist)

Coracoid – robust bone that connects the pelvic girdle and sternum

Cranium – cartilaginous or bony structure that protects the brain

Diaphragm – large flat muscle that forms a wall between the thoracic and abdominal cavities in mammals

Dorsal fin – fin located on the back of an animal

Esophagus – tube that carries food from the oral cavity to the stomach

External gill opening – opening through which water leaves the gill chamber

External gills – gills located outside of the body wall

Falciform ligament – thin membrane that connects the liver of a salamander to the body wall

Feather – hyper developed scale in birds, modified to carry out a variety of functions including insulation, streamlining and flight.

Femur – large bone of the upper hindlimb

Fibula – bone of lower hindlimb

Fin spine – hard, sharp spine located at the leading edge of fins in some fishes

Fin ray – stiff but flexible and soft supporting rods in fins of some fishes

Gall bladder – organ that stores and concentrates bile before it is sent to the small intestine

Gill arches – Stiff support structures that support soft tissue of gills in fishes

Gizzard – muscular organ of the digestive system that grinds up food

Glandular skin – skin that secretes various substances, including water, toxins, mucus, etc., found in amphibians

Gular fold – fold in the skin located in the neck of some animals

Hair – protein filaments produced in follicles in the dermis of mammals

Heart – muscular structure that pumps blood through a circulatory system

Heterodont dentition – dentition that includes teeth of different shapes that carry out different functions, e.g., cutting, tearing, slicing, chewing.

Homodont dentition – dentition that includes teeth that are all the same shape

Hox **genes** – developmental regulatory genes of animals, these genes determine things like body symmetry as well as the timing of activating and de-activating developmental genes

Humerus – large bone of the upper forearm

Kidney – organ that produces urine and contributes to osmoregulation

Large intestine – portion of the digestive tract that carries water absorption

Lateral line – organ embedded along the flanks of a fish, allows the fish to sense vibrations

Liver – large organ that produces bile and is involved in various metabolic processes including metabolism and short-term storage of various nutrients

Lobe-fin – fin in which flesh, including muscle tissue, extends into the fin itself

Lumbar vertebrae – vertebrae between the thoracic vertebrae and sacral vertebrae

Lung – respiratory organ where gas exchange takes place between the air and the blood

Mammary gland – mammalian organ that produces milk

Mandible – lower jaw

Meckel's cartilage – structure from which jaws of vertebrates evolved

Metacarpals – bones between the wrist and the fingers

Metatarsals – bones between the ankle and the toes

Nostril/Nare – olfactory organ (sense of smell)

Notochord – Stiff but flexible cord made of cartilage-like material, it's structure is a series of small discs stacked anterior to posterior in a supporting tube, the distance between discs limits the range of flexibility

Operculum – bony covering of the gill chamber in some fishes

Oviduct – tube that carries eggs away from the ovary

Palatoquadrate cartilage – Structure that forms part of the upper jaw in a chondrocranium

Pancreas – organ that secretes digestive enzymes and sends them to the small intestine

Pectoral fin – anterior paired fin, supported by the pectoral girdle

Pectoral girdle – skeletal structure that supports the paired pectoral appendages

Pectoral limb – same as the forelimb

Pelvic fin – posterior paired fin, supported by the pelvic girdle

Pelvic girdle – skeletal structure that supports the paired pelvic appendages

Pelvic limb – same as the hindlimb

Pelvis – part of the pelvic girdle that connects the hindlimbs and backbone (hip)

Pericardial sac – fluid-filled sac that houses the heart

Phalanges – bones of fingers and toes

Proventriculus – enlargement of the digestive tract anterior to the gizzard in birds

Pygostyle – multiple vertebrae fused at the base of the backbone in birds, it supports tail feathers and the uropygial gland that produces preen oil

Pyloric sphincter – muscular (usually) passageway between the stomach and the small intestine

Radius – bone of the forearm

Rectum – terminal structure in the digestive system, for short-term storage of feces

Rib – structure connected to some vertebrae, used to support body wall muscles and protect the body cavity

Rostrum – alternative term for a snout

Sacral vertebra – vertebrae associated with the pelvic girdle

Scale – flat bony structure used to protect the outer body wall

Scapula – shoulder blade, part of the pectoral girdle

Scapulocoracoid – structure that includes the scapula and coracoid bones, and supports the humerus and connects to the sternum

Small intestine – organ that carries out digestion and absorption of nutrients

Spiracle – opening allowing water to be pulled over the gills of sharks and relations when the mouth is closed or otherwise occupied

Spleen – Removes old red blood cells, stores white blood cells and platelets and may be a site of red blood cell production in some animals

Sternum – breastbone, bone to which ribs attach

Stomach – enlarged sac of the digestive system where food is stored temporarily and where protein digestion begins before food is moved to the intestine

Swim/Air bladder – gas-filled sac that allows fishes to regulate buoyancy. Some fishes use them as a respiratory organ, and is the organ from which lungs evolved

Synsacrum – vertebrae of the pelvis that are fused to each other

Syrinx – voice box in birds

Tarsals – bones of the ankle and upper foot

Testis/testes – primary male organ, produces sperm

Tetrapod – animal with two pairs of legs

Thoracic cavity – space in the body that houses the lungs and pericardial sac

Thoracic vertebrae – vertebrae that bear ribs

Thymus – organ of the immune system that produces T-cells

Tibia – bone of lower hindlimb

Toe pad – enlargements of the toes that provide increased grip and traction

Trachea – upper portion of the respiratory tract in air-breathing animals

Transverse septum – non-muscular wall of tissue separating the pericardial cavity from the main body cavity

Ulna – bone of the forearm

Vas deferens – tube that carries sperm away from the testes

Vent – another term for the anus

Ventricle – muscular chamber that pushes blood out of the heart

Vertebrae – bones of the backbone

Visceral arches – structure through which blood flows to and from external gills

Index

Acorn worm (see Hemichordata)
Actinopterygii 149, 151-152, 155
Amoeba proteus 30
Amoebozoa 29-30
Amphibia 107, 150-152, 157
Angiosperm 93, 104-105, 110
Annelida 53, 56, 58, 67-68, 71, 73-74
Anolis 161-163
Anthocerotophyta 41-42, 44
Anthophyta (Angiosperm)
Archaeplastida 36
Arthropoda 93, 112, 118-119, 129, 131-132, 139
Ascaris 120-123
Ascomycota 81, 83
Aspergillus 87-88
Asteroidea 142
Bacillariophyta 32
Basidiomycota 84-86
Bivalvia 113-114
Brachiopoda 106
Brittle star (see Ophiuroidea)
Brown algae (see Phaeophyta)
Bryophyta 41-42, 45
Bryozoa 106
Caenorhabditis elegans 119
Cat (see Mammalia) 165-166
Caudofoveata 112
Cephalopoda 113-114
Centipede (see Myriapoda)
Cephalochordata 144-146
Ceratium 31-32
Cestoda 106. 108
Chaos diffluens 30
Charophyceae 40, 42
Chelicerata 131
Chiton (see Polyplacophora)
Chlorophyta 36, 40
Chondrichthyes 149, 150, 152-155
Chordata 93, 141, 144-146
Chromalveolata 29-30
Chytridiomycota/Chytrid 81
Ciliata 32
Cladogram 6-11, 14
Clam (see Bivalvia)
Clonorchis sinensis 108-109
Cnidaria 93-94, 98-100
Coelocanth (see Sarcopterygii)
Comb jelly (see Ctenophore)

Compound microscope 15-17, 19, 23, 25-27, 46
Conifer (see Pinophyta) 57-59, 61-63
Coprinus 85-86
Crayfish (see Crustacea) 132-135
Crinoidea 141
Crustacea 131-132
Ctenophora 93-95, 100
Cyanobacteria 29, 57, 82
Cycadophyta 57
Cycliophora 105
Cypress 58-59
Deuteromycota 81, 86-87
Deuterostomia 104, 141
Diatom (see Bacillariophyta) 29-30, 32-33
Dicotyledonae 67
Digenea 106-107
Dinoflagellata/dinoflagellate 29-32
Dinosaur (see Reptilia) 152
Dirofilaria immitis 128
Dissection microscope 17-18, 27
Dogfish, spiny (see *Squalus*) 152-155
Dracunculus medinensis 127
Earthworm (see Oligochaeta) 107, 110-111, 120
Ecdysozoa 104, 118-119, 129, 137
Echinodermata 93, 141-142
Echinoidea 142
Echiura 110
Enterobius vermicularis 126
Entoprocta 105
Equisetopsida 52
Euglena 29, 35
Euglenozoa 35
Excavata 29, 35
Fern (see Monilophyta)
Flatworm (see Platyhelminthes)
Fucus 33-34
Fungi 7-8, 29, 81-90
Gastrotricha 105
Gastropoda 112, 114
Ginkgo balboa 58
Ginkgophyta 58
Gnathostomulida 105
Gnetophyta 58-59
Gordian worm (see Nematomorpha)
Grasshopper (see Hexapoda) 135-137
Gymnosperm 56-59
Hagfish (see Myxini)
Hemichordata 141

Hexapoda 132, 135
Hirudinea 110
Holothuroidea 142
Hornwort (see Anthocerotophyta)
Horsehair worm (see Nematomorpha)
Horseshoe crab (see Chelicerata)
Horseshoe worm (see Phoronida)
Horsetails (see Equisetopsida)
Hydra 98-99
Insect (see Hexapoda)
Jaw worm (see Gnathostomulida)
Jelly fish (see Cnidaria)
Kinorhyncha 118
Lamp shell (see Brachiopoda)
Lamprey (see Petromyzontida)
Lancelet (see Chordata)
Leech (see Hirudinea)
Leptosporangiate fern (see Polypodiopsida)
Lichen 36, 82, 88-90
Liliopsida 67
Liverwort (see Marchantiophyta)
Lizard (see *Anolis*, Reptilia)
Lophotrochozoa 104, 118
Loricifera 118
Lungfish (see Sarcopterygii)
Magnoliophyta 67
Magnoliopsida 67
Mammalia 150, 152, 165-168
Marchantia 43-44
Marchantiophyta 41-42
Micrognathozoa 106
Millipede (see Myriapoda)
Mold 29, 81-82, 86-87
Mollusca 93, 104-105, 112-114
Monilophyta 50
Monocotyledonae 67
Monogenea 106-107
Monoplacophora 112
Moss animals (see Bryozoa)
Mosses (see Bryophyta)
Mud dragon (see Kinorhyncha)
Mudpuppy (see *Necturus*)
Myriapoda 131
Myxini 149-150
Necturus 157-161
Nematoda 5, 93, 118-128
Nematomorpha 118
Nemertea 105
Obelia 98-100
Octopus (see Cephalopoda)
Oligochaeta 110

Onychophora 119, 129-130
Ophioglossales 51-52
Ophiuroidea 142
Opisthokonta 29
Panarthropoda 129
Paramecium caudatum 32
Penicillium 86-87
Penis worm (see Priapulida)
Perca 155-157
Perch (see *Perca*)
Peridinium 31-32
Peripatus 130
Petromyzontida 149-150
Peziza 83-84
Phaeophyta 33
Phoronida 106
Phytoplankton 30, 32, 36
Pig, fetal (see Mammalia) 165-168
Pinophyta 58
Pinus 60
Placozoa 93-94
Planarian (see Turbellarian)
Platyhelminthes 93, 104, 106
Pleurobrachia 94-95
Pogonophora 110
Polychaeta 110
Polyplacophora 112, 114
Polytrichum 46
Polypodiopsida 53
Porifera 93-94, 96
Priapulida 118
Protostomia 104, 118, 141
Psilopsida 51
Psilotum 51
Pterobranch (see Hemichordata)
Ray-finned fish (see Actinopterygii)
Reptilia 150-152, 161-166
Rhizaria 29
Rhizopus 82-83
Rhombozoa 105
Ribbon worm (see Nemertea)
Rotifera 106
Rules of microscopy 19-20
Salamander (see *Necturus*, Amphibia)
Sarcopterygii 149, 151
Scale bars, making 25-27
Scaphopoda 113
Sea cucumber (see Holothuroidea)
Sea gooseberry (see *Pleurobrachia*)
Sea squirt (see Cephalochordata)
Sea urchin (see Echinoidea)

Seastar (see Asteroidea) 141, 142-144, 146
Shark (see Chondrichthyes)
Sipuncula 110
Snail (see Gastropoda)
Solenogastres 112
Spermatophyte 56-57, 71
Sphagnum 46
Spicule worm (see Caudofoveata and Solenogastres)
Sponge (see Porifera)
Squalus 152-156
Stage micrometer 21-24, 26
Starling (see *Sternus*, Reptilia)
Sternus vulgaris 163-166
Taenia 108-109
Tetrapod 151
Tardigrada 119, 129-130
Thallophyta 41-42, 47, 50
Tracheophyta 41-42, 45, 50
Trematoda 106-108, 114
Trichinella spiralis 123-124
Trilobitomorpha/Trilobita 131
Tubatrix aceti 119
Turbellaria 106-109, 111
Tusk shell (see Scaphopoda)
Urochordata 144-145
Velvet worm (see Onychophora)
Vertebrata 144-145, 149, 152
Vestimentifera 110
Volvox 36
Water bear (see Tardigrada)
Wet-mount slides, making 24-25
Whisk fern (see Psilopsida)
White shark (see Chondrichthyes) 154
Wuchereria bancrofti 125, 128
Yeast 81
Zygomycota 81-82, 86

Alan Holyoak earned BS and MS degrees in Zoology with an emphasis in Aquatic Biology at Brigham Young University and a PhD in Biology at the University of California, Santa Cruz, where he focused on the biology of marine invertebrate animals. He taught courses in zoology and aquatic biology at the university level for 30 years and has carried out research or taught field courses at The Friday Harbor Laboratories (University of Washington), the University of Hawaii, The Joseph M. Long Marine Laboratory (University of California, Santa Cruz), Hopkins Marine Station (Stanford University) and the Oregon Institute of Marine Biology (University of Oregon).

Made in the USA
Middletown, DE
03 January 2025